*Microbial ecosystems of Antarctica*

*Studies in Polar Research*

This series of publications reflects the growth of research activity in and about the polar regions, and provides a means of disseminating the results. Coverage is international and interdisciplinary: the books are relatively short and fully illustrated. Most are surveys of the present state of knowledge in a given subject rather than research reports, conference proceedings or collected papers. The scope of the series is wide and includes studies in all the biological, physical and social sciences.

*Editorial Board*

Dr. A. Clarke, British Antarctic Survey, Cambridge
D. J. Drewry, British Antarctic Survey, Cambridge
S. W. Greene, Department of Botany, University of Reading
M. A. P. Renouf, Memorial University of Newfoundland
P. Wadhams, Scott Polar Research Institute, Cambridge
D. W. H. Walton, British Antarctic Survey, Cambridge
P. J. Williams, Geotechnical Research Laboratories, Carleton University, Ottawa

*Other titles in this series:*

The Antarctic Circumpolar Ocean
*Sir George Deacon*

The Living Tundra*
*Yu. I. Chernov, transl. D. Love*

Arctic Air Pollution
*edited by B. Stonehouse*

The Antarctic Treaty Regime
*edited by Gillian D. Triggs*

Antarctica: The Next Decade
*edited by Sir Anthony Parsons*

Antarctic Mineral Exploitation
*Francisco Orrego Vicuna*

Transit Management in the Northwest Passage
*edited by C. Lamson and D. Vander Zwaag*

Canada's Arctic Waters in International Law
*Donat Pharand*

Vegetation of the Soviet Polar Deserts
*V. Aleksandrova, transl. D. Love*

Reindeer on South Georgia: The Ecology of an Introduced Population
*N. Leader-Williams*

The Biology of Polar Bryophytes and Lichens
*R. Longton*

* Also available in paperback

# Microbial ecosystems
# of Antarctica

WARWICK F. VINCENT

*Taupo Research Laboratory,*
*Division of Marine and Freshwater Science,*
*Department of Scientific and Industrial Research,*
*New Zealand.*

The right of the
University of Cambridge
to print and sell
all manner of books
was granted by
Henry VIII in 1534.
The University has printed
and published continuously
since 1584.

CAMBRIDGE UNIVERSITY PRESS

*Cambridge*

*New York   New Rochelle*

*Melbourne   Sydney*

PUBLISHED BY THE PRESS SYNDICATE OF THE UNIVERSITY OF CAMBRIDGE
The Pitt Building, Trumpington ·Street, Cambridge, United Kingdom

CAMBRIDGE UNIVERSITY PRESS
The Edinburgh Building, Cambridge CB2 2RU, UK
40 West 20th Street, New York NY 10011–4211, USA
477 Williamstown Road, Port Melbourne, VIC 3207, Australia
Ruiz de Alarcón 13, 28014 Madrid, Spain
Dock House, The Waterfront, Cape Town 8001, South Africa

http://www.cambridge.org

First published 1988
First paperback edition 2003

*A catalogue record for this book is available from the British Library*

*Library of Congress cataloguing in publication data*
Vincent, Warwick F.
    Microbial ecosystems of Antarctica.
    (Studies in polar research)
    Includes index.
    1. Microbial ecology—Antarctic regions.
I. Title.  II. Series.
QR100.V56  1988  576´.15´09989  88-7316

ISBN 0 521 32875 6 hardback
ISBN 0 521 54413 0 paperback

Cover: Scanning electron micrograph of a cyanobacterial mat from Fryxell
Stream, southern Victoria Land

# Contents

Contents

# *Preface*

Popular descriptions of antarctic ecology often centre around the penguins, seals and other macrofauna of the Southern Ocean. These animal populations contribute to the unique biology of the south polar region, but throughout much of Antarctica, ecosystem processes are dominated by a very different group of organisms. Large animals and also higher plants are absent or only poorly represented in many antarctic habitats; however a wide range of microscopic lifeforms including protozoa, fungi, bacteria and microalgae are often abundant, and these species interact to form dynamic and sometimes highly structured communities.

To the careful observer Antarctica provides continuous reminders of these microbial worlds: brown-stained ice floating on the ocean, red-coloured snowfields, turbid green waters of the inshore sea and tide cracks, black crusts on the soil, blue-green layers in translucent rock. In the more extreme of these environments micro-organisms constitute the entire community biomass, but even in the Southern Ocean and in coastal moss beds microscopic species play major roles in the biological transfer of energy and materials.

The last two decades have seen enormous advances in our knowledge of the antarctic environment, and of the microbial dynamics throughout this region of the biosphere. New micro-organisms have been discovered, including freeze-tolerant phototrophs and heterotrophs that may be endemic to Antarctica. New biotic environments have been described such as rock crystal habitats, tidal lakes, ice-shelf pools, hypersaline soils, fellfield microhabitats and glacial meltwater streams, that provide unusual opportunities for microbial growth. Other environments and their microbiota, for example the sea-ice ecosystems have become much more clearly defined. Classic perceptions such as 'the highly productive Southern Ocean' have had to be substantially revised as investigations of microalgal productivity and microbial trophic structure have become more detailed and have ranged over wider regions of the south polar ocean. Important new foci of environmental research have also emerged, such as the dynamics of the marginal-ice zone at the edge of the receding sea ice, and the increasing influence of humans particularly in the limited areas of ice-free land.

The aim of this volume is to bring together in a structured format a current understanding of antarctic environments and the microbial (including algal and protozoan) communities that exist within them. The book is designed for those with an interest in the ecology of low temperature and polar environments; for arctic and antarctic scientists who wish to extend their environmental knowledge beyond their speciality discipline; and for microbial ecologists and others who are similarly fascinated by the remarkable microscopic life that can endure or even thrive within extreme environments.

This book attempts to summarise the diverse range of ecosystems throughout the south polar region, the major features of the chemical and physical environment in each type of habitat, and the influence of these features on the population structure and dynamics of the microbiota. Each chapter that is devoted to a specific category of antarctic environment concludes with a brief sketch of the overall trophic structure of the ecosystem; these sections aim to provide the reader with an overview of the carbon, nutrient and energy flows throughout that environment and to introduce some of the larger organisms that ultimately depend upon or interact with the microbial community. A compilation of regional climatic data and general environmental summaries are presented in the appendices to support some of the observations made in the text, and as reference sources for investigators with a working interest in antarctic environmental science. A glossary is provided at the end of the book to allow the non-microbiologist to follow most of the material presented in each chapter.

I gratefully acknowledge the assistance and cooperation I have had from individuals and institutions who have helped me throughout this venture. I especially thank the library staffs of the New Zealand Oceanographic Institute, DSIR (Wellington, New Zealand), Antarctic Division, DSIR (Christchurch, New Zealand), Antarctic Division, Department of Science (Hobart, Australia) and British Antarctic Survey (Cambridge, England). I thank my various antarctic colleagues for their encouragement and helpful comments on sections of the manuscript, especially Paul Broady (University of Canterbury, Christchurch), Anna Palmisano (NASA–Ames, California), Harvey Marchant (Antarctic Division, Hobart),

Robert Wharton (Desert Research Institute, Nevada), Tom Clarkson and Mark Sinclair (New Zealand Meteorological Service, Wellington), Imre Friedmann (Florida State University, Tallahassee), Gordon Fogg (University of North Wales, Menai Bridge); I also thank Professor Fogg for his comments on the temperature–diffusion relationship, Cornelius Sullivan (University of Southern California, Los Angeles), Ron Heath (New Zealand Oceanographic Institute, Wellington), Ian Bayly (Monash University, Melbourne) and Clive Howard-Williams (Taupo Research Laboratory), and also those who have kindly provided illustrative material and who are individually acknowledged in the figure legends. I am entirely responsible, however, for errors or omissions.

I additionally thank the staff of Antarctic Division (DSIR) for their excellent logistics and planning support throughout my involvement in the New Zealand Antarctic Research Programme, the original Directorate of the Division of Marine and Freshwater Science (DSIR) for encouraging the inception of this book, Bettie Souter for assistance with some of the scanning electron micrographs, Janet Simmiss for typing assistance with many of the sections, Gillian Wratt for expert field assistance as well as for patiently lending scale to several of my photographs reproduced here (she is 1.7 m tall), and Ron Heath, the Director of the Division of Marine and Freshwater Science (DSIR) for his support. Finally I thank my wife Connie for her participation with me in antarctic research and for her understanding and encouragement; I dedicate this volume to her.

Warwick F. Vincent
1987

# 1

## Introduction

Extreme cold has shaped the antarctic environment and the microbial communities that live within it. Even throughout summer, air temperatures at the margins of the continent and in the maritime zone lie between $-10$ and $+5\,°C$. The dramatic and discontinuous change in the physical properties of water over this range, in particular at its freezing point (Fig. 1.1), has a far-reaching impact on the chemical and physical characteristics of all potential habitats throughout the region. These environmental effects restrict the types of organisms that can be supported, and severely limit the timing and intensity of biological processes.

The freezing and melt cycle exerts the dominant influence on antarctic life forms, although not always directly. Persistent snow and ice are a feature of many antarctic environments (most obviously the continental ice sheet, and the vast expanse of sea ice that encircles the continent each year), even during the period of maximum biological activity. Freezing and melting dictate the areal extent of these environments that in turn modify other properties of the region – for example the distribution of water masses in the Southern Ocean (see Chapters 4, 5) and the regional variations in climate (Appendix 1). For the organisms which live within snow and glacier ice (see Chapter 2), the various forms of sea ice (see Chapter 3), or the permafrost-influenced soils (see Chapter 8), growth and reproduction are totally dependent upon the minute pockets of liquid water, and during freezing and melting these are subject to major fluctuations in volume, osmolarity, pH and temperature. Ice formation can physically disrupt environments (e.g. parts of the marine benthos subject to ice-scouring or anchor ice effects (see Chapter 6)) or cause large-scale changes in the salinity and gas content of the remaining water and the amount of light that penetrates down into it (e.g. ice-capped lakes and pools (see Chapter 7)). Rapid shifts backwards and forwards across the freezing point may effectively sterilise certain environments or force the

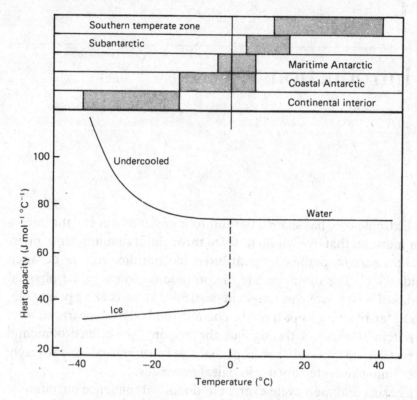

Fig. 1.1. Even in January, the warmest month for biological activity, air temperatures throughout much of Antarctica lie in the range −10 to +5 °C (the shaded bars represent the temperature range for each region, from mean daily minima to mean daily maxima; full data in Appendix 1). The abrupt shift in the properties of $H_2O$ across this range, illustrated here by heat capacity (after Franks, 1985), has a major influence on the distribution and activity of the biota.

micro-organisms into thermally more stable parts of the habitat, e.g. abiotic rock faces and the well developed microbial communities at depth beneath the rock surface (see Chapter 9). Conversely, for certain habitats the formation of an ice cover provides insulating protection from the outside environment, and may allow liquid water to persist despite sub-zero air temperatures.

The subsequent melting process often occurs rapidly. This causes an abrupt improvement in growth conditions that may amplify the effects of the strongly seasonal light regime on primary production and subsequent food chain processes. Most of the Southern Ocean for example, contains sparse concentrations of microalgae (phytoplankton) that have low photo-synthetic rates (see Chapter 5). When the sea ice breaks up each year, the lower salinity meltwater restricts vertical mixing and retains the phyto-plankton under improved light conditions for photosynthesis; this is believed to cause a band of intense primary and secondary production that sweeps across the ocean behind the retreating ice pack (see Chapter 4). Ice shelf (see Chapter 2) and meltwater stream (see Chapter 7) ecosystems in Antarctica are fundamentally linked to radiation and temperature, and small shifts in the energy balance at their ice faces may markedly alter the availability of water for aquatic life in these ecosystems. With summer air temperatures typically remaining close to 0 °C in many antarctic habitats, small changes in local climate can completely shift the balance between freezing and melting and thereby dramatically change the chemical and physical environment. As a result many antarctic ecosystems might be expected to show large interannual variations in the timing and duration of conditions favourable for biological processes.

Marked changes also take place in the immediate cellular environment during the freezing and melt cycle. Extracellular ice formation results in severe dehydration, often compounded by osmotic stress; intracellular ice formation generally causes the death of the cell. Antarctic micro-organisms, like those in other freezing habitats have a range of adaptive strategies that allow them to avoid or at least minimise these destructive effects (see Chapter 10).

Cold temperatures above freezing may also exert a decisive influence on the activity and species composition of antarctic communities. At a cellular level these effects operate in part through the Arrhenius relationship between chemical reaction rate and temperature. However for many species in Antarctica and elsewhere there are major departures from this simple function, even within the ambient temperature range. Some of these responses to cold operate at the cellular macromolecule level. The extent of hydrogen-bonding of proteins with water molecules is a determin-ant of their catalytic activity and this surrounding water structure is sensitive to temperature. Similarly, temperature has a major influence on the fluidity of lipid membranes that in turn regulates many cellular functions (see Chapter 10). These biochemical and biophysical stresses may be exacerbated by further chemical interactions in habitats where freezing has generated high solute concentrations and unusual pH regimes.

Low temperatures may have a differential effect on select components of the community. In microbial consortia of phototrophs and heterotrophs, the markedly different responses to temperature by each functional com-ponent (e.g. fungal respiration versus net photosynthesis by the algae in the microscopic lichens that live within certain rocks (see Chapter 5)) can further generate pronounced non-linearities in the temperature response curve. These effects of low temperature may strongly interact with other environmental variables including light intensity (e.g. net carbon fixation by the rock microbial communities), daylength (e.g. net photosynthesis in the Southern Ocean) and pH (e.g. methanogenesis in moss tundra soils). They may also influence the usefulness of traditional descriptors of microbial activity – for example chlorophyll *a* does not seem to decompose completely in various cold water environments and may be a misleading guide to phototrophic biomass in communities such as the marine benthos (see Chapter 6) and stream microbial mats (see Chapter 7).

Antarctic microbial communities are fiercely regulated by their extreme environment, but they may also have a reciprocal influence on the physical and chemical properties of their surroundings. For example, the presence of dense growths of diatoms in the sea ice substantially lessens its physical strength (Buinitsky, 1977) and may thereby hasten its decay. The communities within the rocks can accelerate the rate of erosion by exfoliation; these rock-dwelling microbes appear to live in a 'precarious equilibrium' between slow growth and destructive biotic and abiotic weathering (Friedmann & Weed, 1987). Similarly the growth of snow algae and ice mats can markedly alter the albedo of these environments and accelerate the rate of ablation. Dense populations of planktonic microalgae within ice-covered lakes can affect the distribution of heat and encourage relatively fast rates of mixing by penetrative convection (Matthews & Heaney, 1987).

Throughout Antarctica micro-organisms such as algae, bacteria, fungi and protozoa play a leading role in the biological transfer of materials and energy. In the extreme portions of this global region, particularly those experiencing near-continuous freezing or very rapid freeze–thaw cycles, higher plants and animals are completely absent and the food chains and biochemical nutrient cycles are entirely microbial. In many antarctic environments, however, human beings and the materials and microbes that we bring with us have generated new, additional tests of microbial resilience (see Chapter 11).

Perhaps the greatest challenge that microscopic communities in Antarctica may face will be the environmental shifts induced by perturbations of the Earth's atmosphere. Current climatic models that examine the effects of man-induced increases in atmospheric $CO_2$ and other gases predict a 0.5–2 °C warming of the temperate zone, and a more pronounced warming in the polar regions, by the middle of the next century (e.g. Schlesinger, 1986). An increase of the ground level temperatures in Antarctica by only a few degrees Celsius would alter the seasonal balance of freezing and melting, with potentially greater ecological repercussions than elsewhere in the biosphere. Human influences on the ozone layer and the resultant increase in ultraviolet radiation particularly in the Antarctic (e.g. Farman, Gardiner & Shanklin, 1985) may also induce strong responses at the microbial level, and cause further shifts in genetic diversity.

True endemism appears to be relatively rare amongst the antarctic microbiota, but the environments of this region have selected for a wide range of robust microbial assemblages. The following chapters explore the structure of these assemblages, the chemical and physical properties of their surroundings and the interactions at both a population and cellular level between the microbiota and their antarctic habitats.

# 2

# Snow and ice ecosystems

## 2.1 Introduction

Snow and ice are dominant features of the antarctic environment. About 12 million km² of land are covered by the continental ice sheet and even the small 'oases' of exposed rock and soil contain alpine glaciers and lake ice, and are periodically covered by snowfalls. The ecology of the marine environment is also greatly influenced by snow and ice. An additional 1.5 million km² of ice extend out over the sea as ice shelves (see Appendix 2), and 17 million km² of the Southern Ocean are covered each year by thick sea ice.

These vast, mostly frozen expanses harbour a relatively small range of suitable habitats for microbial growth. The sea ice provides the most favourable of these environments, and is dealt with separately in the next chapter. During the brief antarctic summer, meltwater is produced at certain locations on non-marine snow and ice. Populations of brightly-coloured snow algae grow within melting snowbanks, particularly in the maritime Antarctic. Small meltholes develop on the surface of glaciers throughout the region and briefly fill with water to support aquatic life each summer. Similar but much larger systems melt out each year over parts of the major ice shelves to form anastomosing networks of pools and streams. Meltpools also form within the ice as large englacial cavities just below the upper surface of the shelf, or as gas- and water-filled bubbles within freshwater lake ice. All of these environments offer potential habitats for microbial colonisation, but each is also characterised by a precarious balance between freezing and melting that may rapidly shift and thereby abruptly alter the conditions for growth.

## 2.2 The environment
### 2.2.1 Snow environments

Antarctic snow is generally deep frozen and dry and therefore provides little opportunity for microbial growth. Colonisation of snowbanks is most likely to occur in the warmer parts of Antarctica such as the maritime zone and at sites within or adjacent to ice-free land where liquid water persists within the snow during summer. Windborne sediment derived from exposed beaches or soil will also be more common in such areas and this material may contain an important microbial inoculum, as well as hasten the melting process by absorbing radiation.

As snow packs melt they appear to become increasingly 'dirty' and clad with sediment (e.g. Warren-Wilson, 1958). In part this may simply result from the cumulative trapping of wind-derived material by the moist surface, coupled with the evaporative concentration of any snow-bound sediment and its accumulation towards the centre of the snowpack by shrinkage (Fig. 2.1). Recent evidence, however, also implicates the electrostatic properties of snow (Benninghoff & Benninghoff, 1985). Wind-blown snow and dust tend to be positively charged and will be deposited in greater quantities on negatively charged (grounded or earthed) surfaces. Thus melting snowbanks in contact with moist soil or bedrock will be more effective in trapping airborne material than dry snow, or even melting snow over ice or permafrost. These electrostatic effects have been demonstrated experimentally near McMurdo Station on Ross Island (Benninghoff and Benninghoff, 1985). The atmospheric potential gradient averages 130 V m$^{-1}$ at sea level in fair weather, but positive gradients up to 5000 V m$^{-1}$ were characteristic of blizzards and periods of blowing snow at McMurdo. Grounded sediment traps under these conditions accumulated up to four times as much snow and dust as non-grounded traps.

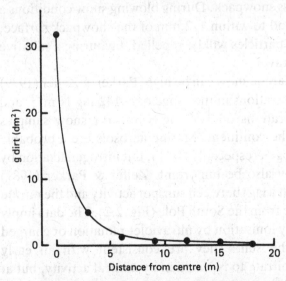

Fig. 2.1. The mass of dirt on a melting snow patch as a function of distance from its centre. Redrawn from Warren-Wilson (1958).

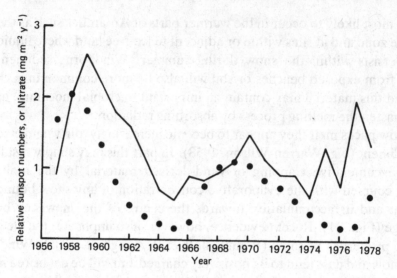

Fig. 2.2. Nitrate in annual layers of snow and ice at the South Pole. The concentrations correlate (at a lag) with relative sunspot numbers (closed circles). Redrawn from Zeller & Parker (1981).

During episodes of falling snow the potential gradient became negative (to $-2500$ V m$^{-1}$) and the grounded traps collected only half the material of non-grounded traps. The airborne sediment contained various microbial fragments, including *Streptococcus*-like cells. The authors suggest that the positive charge on dust and snow may also contribute to the long-distance dispersal of materials across snowpack. During blowing snow conditions a positive gradient may extend to within 1–2 mm of the snowpack surface, and the positively charged particles will be repelled, favouring a saltative or rebounding method of travel.

The nutrient levels in antarctic snow can be high. Parker & Zeller (1979) report mean nitrate concentrations in the range of 5–142 mg N m$^{-3}$, and mean ammonium concentrations of 6–59 mg N m$^{-3}$ for snow samples from various sites across the continent. Marine aerosols are a probable source of much of this nitrogen (especially $NH_4^+$), but nitrogen fixation by atmospheric processes may also be important. Zeller & Parker (1981) report a close correlation (at a lag) between sunspot activity and the nitrate levels in a snow and ice core from the South Pole (Fig. 2.2). The data imply that the nitrate is formed by ionisation by ultraviolet radiation or charged particle emissions from the sun. They are consistent with an early hypothesis attributing the nitrate to $N_2$-fixation by auroral activity, but at present the observations are inconclusive.

Whatever the source, some 27 000 tonnes of nitrate-N and 19 000 tonnes of ammonium-N are deposited on the antarctic ice sheet each year. If the annual snow accumulation approximates the annual losses through calving of icebergs, blowing snow and the melting of the bottom of ice shelves then this nitrogen could be a significant local input to the inshore Southern Ocean (Parker *et al.*, 1978), although it represents only 1–2% of the total dissolved inorganic nitrogen contained within the pelagic waters south of the Polar Front (Biggs, 1978). The melting of the undersurface of the Ross Ice Shelf, for example, transfers a large amount of ammonium that was derived from precipitation over the ice shelf and its inflowing glaciers. This $NH_4$–N may then enter the Ross Sea as a tongue of *shallow ice shelf water* that could potentially influence the nitrogen dynamics of the marine phytoplankton (see Chapter 4).

### 2.2.2  Glacier meltholes – the cryoconite environment

Cryoconite refers to the 'cold rock dust' that blows onto ice sheets and initiates local melting. The term was first used by the arctic explorer Nordenskjold to describe the sediment he found at the bottom of holes over the Greenland Ice Cap (see the excellent review by Wharton *et al.*, 1985). This wind-blown material absorbs heat and melts the underlying ice to form cylindrical or D-shaped holes with near-vertical sides (Fig. 2.3). These depressions commonly form in the *ablation* zone of glaciers (ablation refers to any process that removes snow and ice, such as melting or evaporation) and often contain a rich microflora dominated by algae and cyanobacteria. These micro-organisms must experience low temperatures, periodic freezing and thawing, and a bright light regime enhanced by reflection off the ice surface.

Once a melthole is initiated a number of positive feedback effects accelerate its rate of development. The hole will act as a sediment trap, and will also accumulate the warm (relative to the ice) water flowing over the glacier surface. Water has a low albedo and a higher heat capacity by comparison with ice (see Appendix 2) and so these water-filled holes will absorb more radiation, which in turn will cause faster rates of melting. Microbial communities grow as mats and films within the holes and these may also accelerate ice melt by absorbing solar radiation. Cryoconite holes typically extend up to a few tens of centimetres deep and wide, but they can eventually become connected by surface or subsurface streams, and may expand into large pools several metres across. This process of hole and pond formation may be an important ablation process for glaciers and may accelerate their rate of ice melting and wastage.

Fig. 2.3. A cryoconite hole on the Canada Glacier (Taylor Valley, Southern Victoria Land). Cryoconite refers to the sediment material that accelerates melting and accumulates in the bottom of the hole (below).

Two processes slow down the deepening of cryoconite holes. Radiation reaches the bottom cryoconite layer mostly as visible light that is transmitted obliquely through the ice because of the low solar angles. The amount of energy available for melting will decrease as an exponential function of hole depth (Gribbon, 1979). The second important mechanism of energy transfer to the bottom of the hole is by convection through whatever water it contains, and this process becomes increasingly difficult as the water column deepens. Cryoconite holes therefore tend to deepen rapidly and then remain at a relatively constant depth that is determined by the local climatological conditions. Most cryoconite holes in a specific area have a similar depth, and conversely this average depth may be a reasonable indicator of local climate (Wharton *et al.*, 1985).

### 2.2.3 Ice-shelf systems

The surface of certain ice shelves, large glaciers and ice sheets around Antarctica melt out each summer to form reticulated networks of lakes, pools and streams. These extensive areas of standing and flowing water persist for several weeks, sometimes with a surface crust of ice, before refreezing. This surface flooding is often a feature of ice-covered regions that receive inputs of air-borne sediment which reduce albedo and hasten melting. The runoff from such areas may be a locally important input of freshwater and nutrients to the sea, perhaps enhancing inshore production by stabilising the water column and maintaining phytoplankton in the euphotic zone (see Chapter 4). Three ice-shelf regions, the McMurdo Ice Shelf, George VI Shelf and the Amery Ice Shelf illustrate this fascinating but as yet little explored ecosystem.

The western extension of the Ross Ice Shelf near McMurdo Sound, known as the McMurdo Ice Shelf, has a complex topography of pressure ridges, ablated ice pinnacles and belts of glacial moraine, and contains a large number of pools and streams that melt out for 1–2 months each year (Fig. 2.4). This ablation area encompasses more than 2000 km$^2$ with up to 60% coverage by shallow open water flowing over ice and moraine. Some of the remarkable features of this unusual icescape have been captured by Swithinbank (1970):

> There are few [other glaciers] where moraines exposed on the moving ice may span ten thousand years of the glacier's history; fewer still where beautifully preserved organic remains await the geologist with his dating methods. There are few where rushing torrents course by dazzling ice pinnacles, and where smooth snows and ice pingos are found close to each other.

The streams are narrow (2–20 m wide), but up to several tens of kilometres long, and near land they flow between pressure ridges running parallel to the coast (Brady & Batts, 1981). The main portion of meltwater development contains a large amount of sediment that has been derived from the sea floor at the zones of grounding of the ice shelf (Kellogg & Kellogg, 1984). The sediment forms drift lines that are elevated up to 10 m

Fig. 2.4. Aerial view of the McMurdo Ice Shelf system. The bands of moraine are derived from the sea floor and cause ridges in the ice. Photography by the United States Navy.

above the main surface of the ice shelf, probably due to decreased ablation rates (Fig. 2.4). In general, dirt layers thicker than about 10 mm tend to insulate the ice beneath and retard ablation (Fig. 2.5), thereby creating local mounds or ridges.

The sediments coating the surface of the McMurdo Ice Shelf are also derived at ungrounded sites by *anchor ice*. These buoyant ice crystals form around objects on the sea floor and then float them upwards to beneath the ice shelf. The material is incorporated into the ice by basal freezing, and gradually moves to the surface by ablation. Shells, bryozoa and siliceous sponges are found at many sites on the ice shelf and are believed to have been deposited by this mechanism.

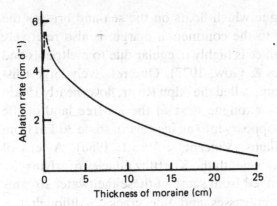

Fig. 2.5. Ablation rate (in centimetres of ice lost per day) as a function of the thickness of overlying moraine material. Redrawn from Ostrem (1959).

Meltwater pools and streams are also found, albeit less commonly, over eastern portions of the McMurdo Ice Shelf that are relatively free of sediment. For example, subsurface or *englacial* pools were discovered during the construction of an aircraft runway on the Ice Shelf about 14 km south of McMurdo Station. These pools were typically one metre deep and spanned circular areas of 10–20 m in diameter. Subsurface melting began in mid-December at a depth of 30 cm or more and progressed until refreezing started in late January. Later in the season the ice cover decreased to 75 mm or less. These pools presented a serious hazard to aircraft traffic and a difficult obstacle for road construction. The melting only occurred beneath blue ice which had a much lower albedo (48%) relative to freshly fallen snow in the area (86%). Only a small shift in the radiation balance was required to inhibit this subsurface melting: a 30 mm thick layer of snow or pulverised ice (albedo of 76%) completely prevented the formation of meltpools (Paige, 1968).

When the McMurdo subsurface pools refreeze in late summer the enclosed water is trapped under increasing pressure. This hydrostatic pressure is relieved when the roof of the pool is uplifted to form an ice pingo, or by release of the water through surface cracks during freezing. These winter ice domes are also found on various lakes and ponds elsewhere in Antarctica including some of the large cryoconite pools.

Further eastwards into the main portion of the Ross Ice Shelf snow accumulation exceeds ablation and there are neither surface nor subsurface meltpools. The McMurdo Ice Shelf therefore spans an interesting transitional climatic zone with increasingly large evaporative losses towards the west.

The Koettlitz Glacier tongue which floats on the sea and bridges the western McMurdo Ice Shelf to the continental margin is also relatively clean of sediment but the surface is highly irregular due to meltpools and stream channels (e.g. Kovacs & Gow, 1977). One relatively large (discharge of about 1 $m^3s^{-1}$) stream, called the Alph River, flows nearly 15 km over the surface of this glacier tongue next to the ice-free land of the continent, and eventually disappears into an ice cavern some 30 km from the open sea (Howard-Williams, Vincent & Wratt, 1986). A lens of freshwater has been identified beneath the Koettlitz Glacier overlying the seawater, and is probably derived from some of these meltwater streams draining down through the crevasses and tide cracks. Although this subsurface meltwater is not saline it isotopically resembles seawater ($^2H$ and $^{18}O$ abundance). It appears that the lower section of the Koettlitz Ice Tongue is primarily formed by seawater that is desalinated during the freezing process, and the isotopic chemistry of its surface meltwaters reflects this marine origin (Gow & Epstein, 1972). Further up the Koettlitz Glacier there is a change in ice crystal structure indicating its freshwater origins in the upper reaches (Fig. 2.6).

The physical and chemical properties of the McMurdo Ice Shelf meltwaters vary enormously. The pools range from freshwater (less than 0.1‰) to more than three times saltier than seawater (up to 100‰). Brady (1980) reports pH values from 6 to 10, and water temperatures up to +7°C while air temperatures were below freezing. The waters appear to have relatively high concentrations of dissolved organic material (up to 950 mg $m^{-3}$ of dissolved organic nitrogen, W. F. Vincent & C. Howard-Williams, unpublished), but a highly variable content of inorganic nitrogen and phosphorus.

On parts of the McMurdo Ice Shelf there are large deposits of mirabilite ($Na_2SO_4 \cdot 10H_2O$) which form salt beds up to one metre thick and

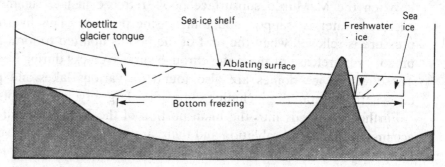

Fig. 2.6. Cross section of the Koettlitz glacier tongue showing the major ice types and formation processes. Redrawn from Gow & Epstein (1972).

36 000 m$^2$ in area (Brady & Batts, 1981). The beds are thought to be derived from seawater brines formed during the basal freezing of the ice shelf (Debenham, 1920). As the shelf grounded these basal brines would have been displaced to the surface through crevasses or tide cracks in the ice and the salts precipitated during subsequent freezing; alternatively the salt crystals may form in trapped basins of seawater beneath the ice shelf during freezing and then gradually travel up through the ice by surface ablation and basal freezing. Mirabilite can be experimentally induced to precipitate from seawater, and forms after 88% of the water is frozen between the temperatures of −8.9 and −22.9 °C (Nelson & Thompson, 1954). These large salt deposits are overlaid by a thick (up to 26 cm) mat of fossilised algae which have been dated at 870 ± 70 years before present (b.p.). These non-marine algae appear to have bloomed in nutrient-rich pools on the ice sheet after the deposition of the mirabilite (Brady & Batts, 1981).

Extensive flooding is also a feature of George VI Ice Shelf between Alexander Island and the Antarctic Peninsula. These meltpools were originally attributed to dust contamination (Wager, 1972) which lowers the albedo of the snow and ice and thereby increases the melting rate. Even dark powder spread on a snow surface as thinly as 1 g m$^{-2}$ can increase absorbed radiation by 50% (Megahan, Meiman & Goodell, 1970), but other explanations of meltwater accumulation in this region have been suggested. Dust contamination is probably important for meltwater formation in the vicinity of Ablation Point and other areas close to exposed moraine and outcrops of fine sediment, but the main lake basins are the result of ice movements and wind action.

Most of the larger lakes on George VI Ice Shelf are long and narrow and are oriented roughly parallel with each other (Fig. 2.7). They are separated by undulations of ice up to 30 m high which are oriented along the direction of glacier flow lines. Another set of much smaller elongate pools are oriented at right angles to the large lakes but parallel with the prevailing wind. These are probably due to the effects of wind action on the ice sheet such as *sastrugi* (wind-eroded corrugations in the ice) which create surface irregularities that channel the water (Reynolds, 1981a).

These sorts of ice basins can only collect meltwater where the snow accumulation rate is low. Reynolds (1981a) suggests a threshold of about $2 \times 10^5$ g of snow m$^{-2}$ y$^{-1}$ below which meltwater pools are to be expected. The accumulation rates are lower than this in the area of lakes on George VI Ice Shelf, but exceed this value elsewhere on the shelf. The Amery Ice Shelf in East Antarctica also develops surface lakes and has a

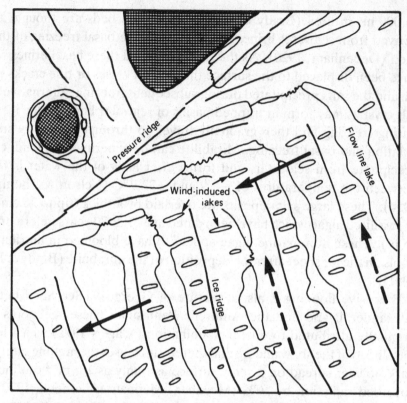

Fig. 2.7. Orientation of lakes on George VI Ice Shelf. The shaded areas represent rock, the solid arrows the prevailing wind direction and the broken arrows the direction of ice flow. Redrawn from Reynolds (1981a).

similar accumulation rate. Air temperatures are comparable over the Amery and George VI Ice Shelves (mean annual temperature of −6––10°C), but are much cooler on the McMurdo Ice Shelf (about −25°C) and may therefore be less important in controlling meltwater formation. Even at a depth of 10 m the George VI Ice Shelf is over 2°C higher than the mean average air temperature, possibly due to the warming effect of refreezing meltwater (Reynolds, 1981a).

Once formed, the larger lakes on George VI Ice Shelf seem to persist for at least several years. The ice surface is very smooth and snow cover is delayed because any precipitation in early autumn is blown clear of the lakes. The low albedo lake ice rapidly absorbs radiation penetrating through the snow in spring, and the snow cover therefore melts faster over lake ice than over *firn* (snow that has been partially consolidated by melting and refreezing).

Like the McMurdo Ice Shelf, the George VI Ice Shelf also contains many meltpools completely contained within the glacial ice and snow. These englacial pools are located within a metre of the snow surface and can be several tens of centimetres deep. At the period of maximum melt an estimated 0.3 km$^3$ of water is contained englacially in the ice shelf while 1–2 km$^3$ are held in surface lakes. Meltstreams drain about 0.4 km$^3$ of water into the marginal zones of the ice shelf each summer, giving a total volume of annual surface melt of about 2.2 ($\pm 0.8$) km$^3$. This is still small (about one-tenth) relative to the water volume estimated to be melting off the bottom of the ice shelf each year (Reynolds, 1981a).

Permeability is also an important feature of the ice shelf surface controlling the extent of meltpool formation. Firn generally becomes impermeable at a density of about $0.82 \times 10^6$ g m$^{-3}$. In the lake area on George VI Ice Shelf firn densities are about $0.86 \times 10^6$ g m$^{-3}$. In other areas on the ice shelf where there has been a high snow accumulation densities are as low as $0.44 \times 10^6$ g m$^{-3}$, and any meltwater formed on this snowpack readily drains away.

Much less is known about the meltwater system on the Amery Ice Shelf. Large surface rivers are developed over wide areas, and meltstreams have also been observed on the inflowing Lambert Glacier 450 km from the sea and at 900 m elevation. Lakes and pools are common, including an *epishelf* lake (Beaver Lake) which consists of freshwater that is dammed by ice but directly overlies seawater and therefore rises and falls with the tides (Mellor & McKinnon, 1960). This tidal influence on lakes and pools will occur wherever they are underlaid by cracks through the ice shelf that allow the underlying seawater to penetrate up to its equilibrium level. Such effects have also been observed in certain parts of the McMurdo Ice Shelf and George VI Ice Shelf. Extensive meltwater features have been reported from other ice shelves in the antarctic region (e.g. Reynolds, 1981b) and are also known from the Arctic (e.g. on the Greenland ice cap, Maurette *et al.*, 1986).

### 2.2.4   *Ice-bubble habitats*

The ice overlying antarctic lakes and pools differs considerably in its chemical and physical properties from the sea ice which extends over much of the Southern Ocean. This freshwater ice lacks the brine channels and loose crystal structure which characterise the bottom-most layers of sea ice that harbour a rich community of marine micro-organisms (see Chapter 3). The paucity of liquid water habitats in lake ice reduces the opportunities for microbial life. However, gas bubbles may become trapped in the ice and gradually move to the surface by the freezing of water beneath and removal of ice from the upper surface by ablation. These ice bubbles act as solar energy traps and develop their own microenvironments of gas and water (Meyer-Rochow, 1979).

Ice-bubble systems are particularly common in the ice overlying lakes with benthic mat communities of cyanobacteria. Pieces of mat become buoyed with oxygen and nitrogen and then break off and float upward to accumulate beneath the ice cap. When the ice cap begins to thicken in autumn these algal mats are incorporated within it. Each mat subsequently absorbs radiation penetrating through the ice and melts the surrounding ice to form a bubble filled with gas and water. This ice bubble ecosystem then gradually moves to the ice surface where the mat is released to the atmosphere. It has been suggested that for some antarctic lakes these ice-bubble systems may represent an important 'escape' mechanism for microbes and nutrients (Wilson, 1965; Parker *et al.*, 1982c). Ice-bubble communities are also found in lakes that remain almost completely frozen to their base throughout the year, e.g. Lake Vida in the Dry Valley region. These lakes presumably contain mats of cyanobacteria in the thin layer of liquid water that is produced by melting at the ice–sediment interface each summer.

## 2.3    Microbial community structure

### 2.3.1    Snow communities

Algae growing in or on the surface of snowbanks and ice are well known from temperate latitudes, but seem relatively rare throughout the antarctic region. Most of the reports of red and green snow communities (e.g. the records compiled in Kol, 1968) are from the maritime zone (western and northeastern parts of the Antarctic Peninsula, and associated islands to the west and north; see Fig. A1.1). Snow algae seem to be puzzlingly absent from melting snowfields at coastal sites elsewhere in the region, for example snowdrifts that persist through summer at Cape Bird, Ross Island (Broady, 1984a). These communities appear only to be conspicuous where the mean January air temperature lies at or above 0 °C (see Appendix 1) suggesting that the slow-growing algae require prolonged periods of melt to build up to a size that produces a visible stain to the snow.

Red-, green- or yellow-coloured snow is commonly encountered on the South Orkney Islands in late summer. Fogg (1967) found that patches of algae developed rapidly during the thaws, and were generally dominated by the green flagellate *Chlamydomonas nivalis* (Sommerf.) Wille, the short-filament chlorophyte *Raphidonema nivale* Lag. and a chrysophyte tentatively identified as *Ochromonas*. The intensely red spores of *C. nivalis* and *Chlorosphaera antarctica* Fritsch often determined the colour of the snow despite large variations in the abundance of vegetative cells of these and other species. The filamentous green alga *Hormidium subtile* was only found in green snow, and *Raphidonema* dominated both the green- and yellow-coloured communities, the latter due to chlorotic cells.

These Orkney Island algal populations were generally located in the uppermost centimetre of snow with very low cell numbers at greater depths. However, green-coloured snow was sometimes detected several centimetres below the surface. After snowfalls the peak concentrations remained at the top of the firn layer beneath the fresh snow, but this layer was eventually exposed after surface melting and ablation (Fig. 2.8). Green, actively motile cells of *Chlamydomonas nivalis* and green cells of *Chlorosphaera antarctica* were recorded at concentrations of $c$. 1 cell $mm^{-3}$ down to a 25 cm depth below patches of red snow. Many samples of firn snow with no obvious coloration contained 1–2 cells $mm^{-3}$ of *Raphidonema* or red spores, indicating that this type of community may be more widely distributed through the antarctic region than sparse records of coloured snow would suggest. These cell concentrations in uncoloured

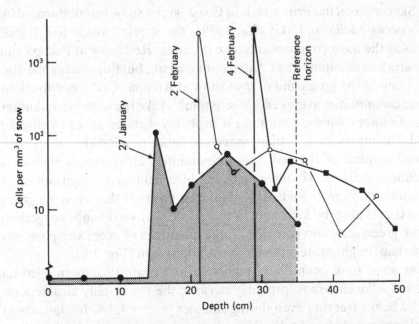

Fig. 2.8. Changes in the depth distribution of snow algae on Signy Island. Redrawn from Fogg (1967).

snow, however, were very low by comparison with population densities in the green and red snow patches, e.g. up to about 5000 cells mm$^{-3}$ for *Raphidonema nivale* in green snow, and 40 spores mm$^{-3}$ in red snow.

Samples from the Orkneys were first examined by Fritsch (1912) who identified 23 taxa of snow algae and established the new chlorococcalean genus *Scotiella*. This genus has been subsequently recorded as one of the most common cryoalgae in alpine regions in the temperate zone, but more recent studies in Antarctica have questioned the validity of this taxon. In the Prince Halard Coast region (*c.* lat. 59 °S, 40 °E) the main constituents of coloured snow were *Scotiella polyptera*, *Cryocystis brevispina*, *Chlamydomonas* sp. and *Stichococcus bacillaris*. Scanning electron micrographs of *S. polyptera* revealed that the cells were frequently coated with a thin membrane that resembled the primary membrane of volvocalean zygospores (Akiyama, 1979). From this and temperate latitude studies it appears that *Scotiella polyptera* is the resting stage of a green alga in the order volvocales, probably a species of *Chloromonas*. It is now known that the life cycles of the volvocalean snow algae are quite complex with morphologically distinct motile and non-motile zygote stages that have previously confused taxonomists. The zygotes from *Chloromonas brevispina*, for example, are identical to seven snow algal taxa previously assigned to the Chlorococcales (Hoham, 1980).

At Skarvsnes on the Prince Halard Coast, green snow was dominated by *Stichococcus bacillaris* (Akiyama, 1979), the species which Kol (1968) regards as the most common antarctic cryoalga. Red snow at Rumpa and Skarvsnes was dominated by *Chlamydomonas* sp., but this species was also a dominant in the green and orange snow at Rumpa. Cell concentrations in these communities averaged 300 cells mm$^{-3}$. Like the snow communities on the Orkney Islands, coloration is probably dictated more by physiological conditions rather than species composition (Akiyama, 1979). Acetone extracts of the red snow communities at Skarvsnes showed a pronounced peak at 475 nm suggesting high cellular concentrations of astaxanthin-like carotenoids. It is of interest however that even the green snow extracts absorbed strongly in the carotenoid wavebands suggesting that the green cells also contained large quantities of accessory pigments that perhaps might protect them against bright light (Fig. 2.9).

Snow algae have been rarely reported from coastal southern Victoria Land and adjacent areas, probably because the mean daily air temperatures lie below freezing even during summer (e.g. −4.9 °C for January at Scott Base, see Appendix 1). Warmer temperatures occur in the inland Dry Valleys (e.g. 1.2 °C for January at Vanda Station), but this area

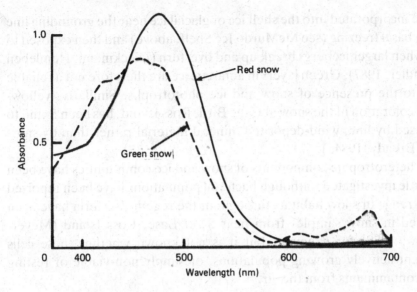

Fig. 2.9. Absorption spectra for acetone extracts of snow algal communities at Skarvsnes. Redrawn from Akiyama (1979).

receives a very small annual snowfall and by mid-summer most of the snow has completely melted away. Further northward on the coast, however, melting snow banks and the associated algal communities are more common.

Green algal layers on and within ice have been recorded from several antarctic sites, but the most detailed description is from the Balleny Islands (67 °S, 163 °E), about 240 km northwest of Cape Adare, northern Victoria Land. 'Pure green layers in the ice' were recorded on the ice cap of one of the islands and a large green icicle was sampled for microscopic examination (Kol & Flint, 1968). The microscopic community was dominated by green algae and contained several new taxa including *Chlamydomonas ballenyana* (the dominant), *Ankistrodesmus antarcticus*, *Bracteacoccus minor* var. *glacialis* and *Ellipsoidion perminimum* var. *cryophila*. A diatom (*Nitzschia* sp.) and cyanobacterium (*Phormidium priestleyi*) were also recorded. This assemblage was probably concentrated from a meltpool or stream, but it is not known to what extent such communities could grow within pockets of water within the ice.

Translucent green ('bottle green') coloured icebergs are sometimes encountered in the Southern Ocean. The reports to date indicate that this coloration is caused by absorption and scattering of light by non-living, mostly inorganic debris rather than by pigmented algae. Such material may be incorporated into the shelf ice or glacial ice near the grounding line during basal freezing (see McMurdo Ice Shelf, above) and then exposed to view when larger icebergs break up and overturn (Dieckmann, Hemleben & Spindler, 1987). Green layers in antarctic ice are therefore not a reliable guide to the presence of snow and ice phototrophs. Similarly, yellow-brown coloration of the snow at Cape Bird, Ross Island, has been found to be caused by fine, wind-deposited mineral material rather than by snow algae (Broady, 1984a).

The heterotrophic components of snow and ice communities have been very little investigated, although bacterial populations have been reported from a range of snow habitats throughout the region. Bacteria have been observed in snow samples from near Scott Base, Ross Island (Meyer-Rochow, 1979) for example, but it is not known whether these cells represent actively growing populations, or simply non-viable or resting spore contaminants from the air.

### 2.3.2    *Cryoconite communities*

The cryoconite habitat supports a much more diverse assemblage of micro-organisms than snow and is typically colonised by cyanobacteria, green algae, diatoms, fungi and bacteria. Two types of cryoconite community have been described in some detail in the McMurdo Sound region.

Cryoconite holes and supraglacial pools are abundant across the wastage area of the Canada Glacier which lies in one of the so-called 'Dry Valleys' of southern Victoria Land (lat. 77° 38'S, long. 162° 53'E, see Appendix 1). These meltwater holes contained macroscopic (up to 2 mm) pieces of microbial mat that were primarily composed of entangled filaments of the cyanobacterium *Phormidium frigidum* Fritsch. Several other cyanobacteria were also present including members of the genera *Microcoleus, Nodularia, Lyngbya, Nostoc* and *Synechococcus*. Green and sometimes orange cells of the chlorophyte *Tetracystis* were embedded in the mat matrix, and occasionally the flagellate *Chlamydomonas subcaudata* Wille and the diatom *Caloneis centricosa* var. *truncatula* (Grunow) Meister were observed. This community differed considerably from those reported on the Greenland ice cap and on North American glaciers, but the species were well known from Dry Valley streams, lakes and soils and are probably derived from these nearby aquatic and terrestrial sources (Wharton *et al.* 1981).

A different species assemblage has been reported in the cryoconite holes on the Mt. Bird Ice Sheet near the northwestern coast of Ross Island (Broady, 1984a). The pools were typically circular, 20–40 cm in diameter and 20–30 cm deep. They were covered by a thin sheet of ice and inside contained water overlying the cryoconite sediment. More than 14 species of algae were observed in this habitat. The dominant taxa were *Phormidium frigidum* and *P. laminosum*, various chlorococcales, the filamentous green alga *Klebsormidium*, an unidentified green alga resembling *Desmococcus*, and two species of *Navicula* (*N. cymatopleura* and *N. muticopsis*). A member of the diatom genus *Achnanthes* was also common in some of the meltholes, including a pool some 5 m² in area.

### 2.3.3  *Ice-shelf communities*

Benthic mats of cyanobacteria and diatoms appear to be the ecosystem dominants in the ice-shelf meltwater systems that have been explored to date. At least two assemblages of diatoms have been distinguished in samples from the McMurdo Ice Shelf system, the first dominated by *Nitzschia westii* and *Pinnularia cymatopleura* (synonymous with *Navicula cymatopleura*), and the second by several species of *Navicula*. *Achanthes tavlorensis* and *Melosira* spp. have also been frequently recorded in samples from this region (Kellogg & Kellogg, 1984), but the environmental factors which control these distributions are completely unknown. Pink and grey benthic mats appear to line the bottom of many of these ice-bound pools (Fig. 2.10), and are dominated by the oscillatoriacean genera *Oscillatoria*, *Lyngbya*, *Phormidium* and *Microcoleus*.

Fig. 2.10. A small meltpool on the McMurdo Ice Shelf showing the benthic film of cyanobacteria.

### 2.3.4   *Ice-bubble communities*

These freshwater ice assemblages are dominated by cyanobacteria which form benthic mats in the shallow-water regions of the lake (see Chapter 7). The dominant is generally *Phormidium frigidum* which forms a 2–10 mm thick cohesive mat over the lake floor with vertical structures up to 5 cm high that are filled with oxygen gas bubbles produced by photosynthesis.

These columnar lift-off mats tear off from the sediments, float up and accumulate beneath the lake ice, and are then incorporated into the ice cap by freezing. The mats form discrete horizontal layers within the lake ice that correspond with the annual freezing cycle; for example the five metre thick ice on Lake Hoare accumulates (and loses by ablation) about 50 cm of ice $y^{-1}$ and contains 6–10 annual layers of mat clumps through the ice column (Parker *et al.*, 1982c). It is not known how much growth takes place during this 10 year traverse of the ice cap, but photosynthesis may accelerate in the final years with increasing melting and improved light availability.

The mats also contain diatoms such as *Hantzschia* and *Navicula*, entrapped sediment and calcite crystals that presumably have been formed biogenically. The organic content of the mats ranged from 8 to 64% (dry weight), with an unusually low ratio of Kjeldahl-N to total phosphorus. The silica content of the mats is high, probably reflecting their inorganic sediment content rather than the presence of diatoms.

The ice-bubble communities absorb radiation and melt the surrounding ice, particularly in the upper 50 cm of the ice cap where an upward-advancing columnar channel is often observed, accelerating the movement of the mat toward the surface (Wilson, 1965). Once at the surface the mat is freeze-dried and dispersed by the wind.

### 2.3.5   *Continental ice-sheet flora*

With its sustained extremes of cold and aridity the Antarctic Ice Sheet provides the harshest environment for microbial life throughout the south polar region. Despite these severe conditions a wide range of viable micro-organisms have been identified or cultured from the snow and ice in central Antarctica. Russian investigators have isolated bacteria, fungi and actinomycetes, including a new actinomycete *Nocardiopsis antarcticus* that synthesises high cellular concentrations of melanin (Abyzov, Filippova & Kuznetsov, 1983). Various bacteria have been observed in snow samples from the surface and one metre depth near South Pole Station (Meyer-Rochow, 1979). There has been no attempt to measure the ATP content or other metabolic properties of these micro-organisms and it remains possible that at least some of the cells are briefly active each year. Most of these microbes, however, are probably inactive and have simply precipitated out from the atmosphere. As noted by Llano (1972), the snow fields of the continent offer an enormous 'natural catchbasin' for atmospheric particles, and given the ability of many micro-organisms to survive freezing and drying the diversity of isolates from the interior is not surprising.

## 2.4    Microbial processes

Investigations of the metabolic properties of snow and ice microflora have been sparse and the production dynamics of most of the communities described above are completely unknown. Fogg (1967) measured photosynthesis by the snow algae of the South Orkney Islands and recorded extremely slow rates. Cellular rates of carbon fixation were less than 5% of the rates measured in exponentially-growing *Chlorella* cultures, and gave a theoretical generation time of about 23 days. *In situ* measurements at a snow temperature of 0 °C gave an areal fixation estimate of about 10 mg C $m^{-2}$ $d^{-1}$, an order of magnitude less than sea-ice communities for example. An increase in temperature from 0 to 15 °C under light-limiting conditions increased photosynthetic carbon fixation by only a few percent.

Snow algae are widely distributed through alpine areas of the temperate zone and much of the information from Northern Hemisphere studies is probably applicable to antarctic communities. These temperate cryoalgae have growth optima below 10 °C and some species seem to have a very narrow tolerance range for temperature. *Chlainomonas rubra*, for example, sheds its flagella at temperatures above 4 °C and suffers membrane damage when frozen at −1 °C (see the review by Hoham, 1980).

## 2.5    Trophic structure

The trophic relationships of non-marine snow and ice communities are extremely simple. The limited range of phototrophs convert solar energy to biomass and are probably slowly and only partially decomposed when they die by bacteria and fungi. In the more aqueous systems such as cryoconite holes and ice-shelf meltpools, the food web is much more developed and includes protozoa, nematodes and rotifers. A similar trophic structure probably characterises the ice-bubble community which for several years may function as a closed microcosm before it reaches the top of the ice cap and is expelled to the external environment.

# 3

# Sea-ice ecosystems

## 3.1    Introduction

Vast areas of ice form and melt each year at the surface of the ocean surrounding Antarctica. This frozen layer grows to 1–5 m thickness and at its maximum extent in August–October forms a continuous, 400–1900 km wide belt that completely encircles the continent. The huge expanse of sea ice persists for several months and provides a remarkably structured and dynamic environment for microbial growth.

There are large regional differences in the physical characteristics of antarctic sea ice, and the species composition of its microflora is correspondingly varied. Three broad categories of microbial community can be distinguished in the ice: surface communities, including meltpool and tide crack micro-organisms; interior communities at depth within the ice; and bottom-ice communities that develop in different types of ice near the ice–seawater interface. The relative contribution of each community to total biomass and production varies greatly at both a local and regional level.

Pennate diatoms are usually the biomass dominants in the sea-ice microbial ecosystem. They live in brine pockets and drainage channels of the ice and grow to maximum abundance in the spring and summer (Oct–Jan). Population densities are often highest in the bottom-most ice strata which are turned visibly brown by the algal bloom. During these periods of peak biomass, chlorophyll $a$ levels can reach several hundred mg m$^{-2}$.

Bacteria and protozoans are also important microbial elements of the annual sea-ice community. Most of the bacteria are free-living rods and cocci, but a separate, morphologically diverse assemblage of bacteria colonises the living microalgae and detrital material contained within the ice. This bacterial secondary production increases through the season and supports a detrital food web through bactiverous zooplankton, including protozoa.

The sea-ice microbial community is an important component of carbon and energy flux in the Southern Ocean. The microalgae contribute new organic carbon, and extend the short polar season of phototrophic growth into late spring and early autumn when phytoplankton photosynthesis is negligible. This primary production supports a diverse array of secondary producers, including the microbial heterotrophs and a sea-ice fauna. When the ice melts each year all of these populations are liberated into the water column beneath, and this inoculum may initiate a new sequence of microbial production in the region known as the marginal ice zone (see Chapter 4).

## 3.2     The environment

As a microbial habitat the sea-ice environment poses a number of problems. Ice formation occurs rapidly and is accompanied by large changes in light regime, carbon supply, nutrient content and salinity. Micro-organisms trapped in the ice during early freezing may ultimately experience temperatures close to the antarctic winter air minimum. Planktonic microalgae in the Southern Ocean must survive several weeks without light for photosynthesis, yet these cells may provide the inoculum for the spring ice–algal bloom. Ice decay is more rapid than its formation, and the environmental changes over this period must be similarly abrupt. The microbial life of the sea-ice ecosystem must contend with sharp vertical gradients in ice structure, light, temperature and saltwater concentration, and these conditions are further varied by large-scale shifts over time and space.

### 3.2.1    Sea-ice cover

Satellite imagery is providing an increasing refined picture of the between and within year variations in antarctic sea ice. Long before remote sensing, however, explorers and oceanographers of the Southern Ocean appreciated the large-scale seasonal fluctuations in ice cover (Fig. 3.1). The sea-ice zone of the Arctic varies by less than 25% through the year (Lewis & Weeks, 1971), but coverage over antarctic waters changes more than 300% between seasons (Treshnikov, 1966). Satellite observations over the Southern Ocean from 1973–5 show a minimum coverage of 3 million km$^2$ in February rising to an annual maximum of 20 million km$^2$

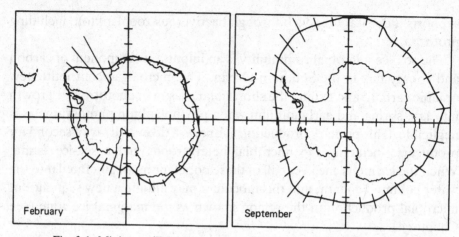

Fig. 3.1. Minimum (February) and maximum (September) sea-ice coverage. The bars represent the range for these months and show the large interannual variation in ice conditions, particularly in summer. Modified from Jacka (1983).

in September–October (Zwally *et al.*, 1979). This places 17 million km² within the sea-ice zone, an area larger than the antarctic continent.

Heavy ice concentrations persist through summer at several locations around Antarctica. The largest sea-ice coverage is in the Weddell Sea west of 45 °E longitude and south of 68 °S latitude (about 0.5 million km²). Sea ice also persists in the Bellingshausen Sea below 70 °S latitude and from the Antarctic Peninsula to 140 °W longitude. Ice conditions can change rapidly (Fig. 3.2). During November–December the northern ice edge retreats at rates up to 16 km d⁻¹ at the Mirny meridian (Treshnikov, 1966). The growth rates of sea-ice cover are less than one half of the decay rate. Most of the decay takes place over the period mid-November to mid-January, decreasing from 17.5 to 6.5 million km². Zwally *et al.* (1979) partitioned the satellite observations into two categories of coverage: greater than 15% of water area covered by sea ice and greater than 85% ice cover (Fig. 3.2). For this second category the sea-ice retreat is spectacularly fast, with a decay from near-maximum to only 25% cover over a four week period between mid-October and mid-November. This rapid breakup and melt cannot be controlled entirely by heat gains from the atmosphere. Gordon (1981) estimates that sea–air heat exchange in the 60 °–70 °S zone can account for only 50% of the required heating and that the remainder is supplied by the upwelling of the relatively warm (about 2 °C) deep water which resides below the Southern Ocean pycnocline. In the Arctic, the sea-ice growth

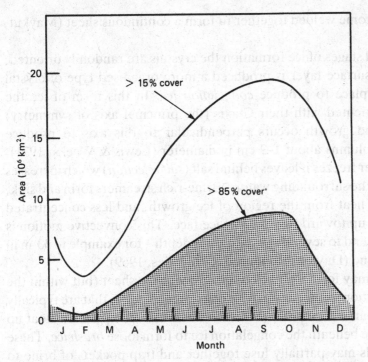

Fig. 3.2. The growth and decay of sea ice around Antarctica. The curves are smoothed from several years satellite data for two categories of ice coverage. Modified from Zwally *et al.* (1979), as cited in Gordon (1981).

rate is much faster than its retreat (Walsh & Johnson, 1979). Gordon (1981) attributes this reverse pattern to the much higher freshwater input into the Arctic Ocean. This induces a strong pycnocline through which the heat flux is relatively small. The ocean–atmosphere heat flux therefore exerts a more controlling influence on annual ice cover in the Arctic than it does in the Antarctic.

### 3.2.2  Ice formation and structure
    Annual sea ice begins to form in the turbulent surface waters of the ocean as small (<3 mm) discs and needles called *frazil ice*. These crystals float up and accumulate in the surface waters to produce a soupy mixture called *grease ice*, but once the ice fraction exceeds about 30–40% the crystals begin to freeze together to form solid sheets of ice 1–10 cm thick. In the presence of wave action the ice cover initially forms as circular pieces 30–300 cm in diameter called *pancake ice*. With continued freezing these

eventually become welded together to form a continuous sheet (Maykut, 1985).

In the initial stages of ice formation the crystals are randomly oriented, but once the surface layer is produced a more organised type of crystal growth takes place to produce *congelation ice*. In this form of ice the crystals are oriented with their C-axis (the principal axis of symmetry) horizontal, and growth occurs perpendicular to this axis to produce interlocking columns about 1–2 cm in diameter (Lewis & Weeks, 1971). As the seawater freezes it leaves behind salt (*salt rejection*) which increases the density of the surrounding waters. Brine-rich streamers form and sink, carrying away heat from the region of ice growth, and less concentrated seawater rises up toward the freezing ice face. This convective motion is believed to extend to several tens of metres depth – for example to 33 m in McMurdo Sound (Dayton, Robilliard & De Vries, 1969).

Ice crystals may form well below the growing ice sheet (but within the zone of convection) either as frazil ice or as large platelets that are typically 10–15 cm in diameter and 0.2–0.3 cm in thickness. The platelets float up and accumulate beneath the congelation ice to form loose *brash ice*. These platelet crystals may partially fuse together and trap pockets of brine to form a porous layer, up to 4 m thick beneath the annual sea ice. Platelets may also form at considerable depth in the Southern Ocean in supercooled water streaming out from under ice shelves; for example, at 250 m depth near the Filchner Ice Shelf (Dieckmann *et al.*, 1986). Platelet ice sometimes freezes to the ocean floor, particularly within the inshore zone of convective circulation, thereby forming *anchor ice* (Dayton *et al.*, 1969). The annual ice sheet can break up and drift to form *pack ice* or may remain attached to the shoreline as *fast ice*.

Cross sections of the annual congelation ice reveal numerous small brine pockets and drainage channels, many less than 100 $\mu$m in diameter (Fig. 2 in Sullivan & Palmisano, 1984). Larger brine-containing chambers are produced during the early stages of freezing when the ice first forms as an open array of interlocking plates. However, these are mostly filled in with congelation ice as freezing continues (Lewis & Weeks, 1971). When big pockets of trapped brine drain down to the ice–water interface the resultant plume can produce an ice stalactite. These structures have been recorded up to 6 m long, at times extending from the sea ice to the ocean floor (Dayton & Martin, 1971).

These various types of ice formation comprise a highly variable percentage of the overall sea-ice structure both between locations in Antarctica and between years. At several sites in McMurdo Sound sampled over three years the seawater–sea-ice interface consisted of a hard congelation ice surface with numerous brine channels in the lowermost 20 cm and many ice stalactites extending one metre or more into the water column. In the 1982 season, however, the two metre thick congelation ice at Cape Armitage had a 25–100 cm thick, loosely consolidated layer of platelet ice with large chambers of interstitial water between the lamellar crystals (Sullivan *et al.*, 1983). A similar type of platelet underlayer was earlier described in McMurdo Sound by Bunt (1963). This loose layer was absent from the hard congelation ice sampled by McConville & Wetherbee (1983) around Casey (66 °S, 100 °E), Davis (68 °S, 78 °E) and Mawson (67 °S, 62 °E). In these regions the sea ice had a firm, relatively flat undersurface. In the Weddell Sea frazil ice seems to form continuously and may comprise the bulk of the annual sea ice – ice floes of the region are primarily composed of the fine-grained equi-axial frazil ice crystals (Gow *et al.*, 1981). These variations in ice structure produce very different types of microenvironment for microbial growth and survival (see Section 3.3).

The relative importance of the different ice types appears to be closely related to the physical environment during ice formation (Weeks & Ackley, 1982). Frazil ice is associated with dynamic, turbulent conditions in the leads (open water between pack ice) and polynynas exposed to wind. The small ice crystals form at fast growth rates ($>1$ cm h$^{-1}$) and can quickly pile up downwind into substantial layers of ice; thicknesses exceeding 50 cm can form by this process within a few hours. By contrast the congelation ice forms by the slow removal of heat from water beneath an existing ice sheet. Once the first few centimetres have formed the low thermal conductivity of ice (see Appendix 2) limits heat transfer and the rate of ice formation slows to $<1$ mm h$^{-1}$. Metre-thick ice requires several weeks to form by this process.

The sea ice is commonly overlain by snow and this layer also may offer opportunities for microbial growth. Snow cover is often much thicker in the Antarctic than in the Arctic, and in regions near the coast that receive blowing snow from the katabatic and other offshore winds this layer may accumulate up to 6 m thick (Keys, 1984). Seawater can penetrate up into the snow, sometimes forming a slush on warm days, but freezing solid during cooler weather (Burkholder & Mandelli, 1965). Where the fast ice is attached to land the tidal rise and fall fractures the ice into a series of parallel *tide cracks*. These accumulate snow which can also become infiltrated with seawater providing a nutrient-rich well-illuminated environment for microalgal growth.

### 3.2.3    *Sea-ice chemistry*

Early data collected by Bunt & Lee (1970) indicated that the McMurdo Sound brash ice contains rich levels of major nutrients. The interstitial waters had 7.07 mmol m$^{-3}$ of nitrate-N and 2.16 mmol m$^{-3}$ of dissolved reactive phosphorus. These values, as well as salinity (31.89‰), pH (8.03) and temperature ($-1.75\,°C$) were comparable with the seawater beneath.

Higher up in the ice column, however, where there is less exchange with the underlying seawater the nutrient levels may be depleted by microbial growth. It is difficult to sample the interstitial water for nutrient analysis but an informative approach has been to melt sections of the ice core and compare the nutrient levels with salinity. In the absence of biological effects nutrient concentrations should follow the dilution curve if the decrease of the nutrient is related solely to reduced salinities in the ice. In Weddell sea ice both silicate and nitrate concentrations fell below the dilution curve, particularly in older ice, indicating nutrient depletion by the ice microbiota. By contrast nitrite values lay above the dilution curve implying an active population of nitrifying bacteria in the sea ice (Clarke & Ackley, 1984).

The salinity of Arctic sea ice is closely related to its age. New rapidly frozen ice contains 20‰ but over the subsequent week of ice accumulation salinity can drop to 7‰. Data from a range of arctic sources compiled by Grainger (1977) show the rapid decrease in salinity with age, and therefore also thickness (Fig. 3.3).

Salinity profiles in McMurdo Sound ice follow the C-shaped pattern that is typical of first year arctic ice (Fig. 3.3): salinity is high near the top, decreases near the middle and then rises again toward the bottom (Weeks & Lee, 1962). This pattern is probably the result of the rapid formation of the near-surface ice trapping larger amounts of salt, the much slower rates of freezing beneath this layer, and the gradual migration of brine as density currents towards the bottom. These concentrated brines may severely stress the microbiota that they come in contact with as they drain down through the ice (see section 10.3).

A reverse pattern is found in the Weddell Sea pack ice where salinity rises to a maximum midway down the ice column. Ackley, Buck & Taguchi (1979) suggest that temperature controls these regional differences. Air temperatures are much higher over the Weddell Sea and may cause some warming of the top few centimetres of the ice sheet. This results in a partial melting and migration of brine thereby enriching the interior of the ice.

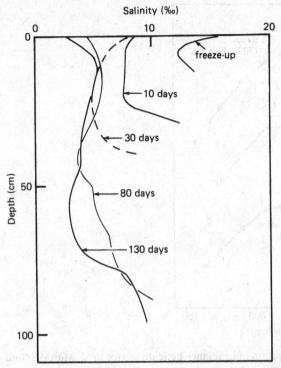

Fig. 3.3. Depth distribution of salinity in first year sea ice in the Arctic. New ice has a high salinity and a C-shaped profile, but brine drainage and slower freezing rates produce more even profiles later in the season. Modified from Grainger (1977).

However, air temperatures do not rise sufficiently to allow complete brine drainage as occurs elsewhere. In support of this hypothesised mechanism, Ackley *et al.* (1979) report that at a more southern station on the fast ice close to the Weddell Ice Sheet, air temperatures dropped considerably (from −2 to −12 °C) and uniformly high salinity was measured throughout the profile.

### 3.2.4 *Light and temperature*

Light and temperature gradients in the sea-ice column are extreme. Near-surface ice must experience antarctic winter air temperatures which fall below −50 °C (see Appendix 1). By contrast, the temperature of the lower ice surface must lie near seawater temperature of about −2 °C.

Light changes in both quantity and quality down the sea-ice column. Two metres of congelation ice reduced surface photosynthetically avail-

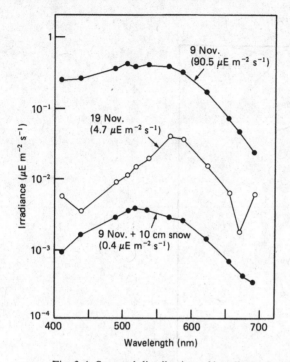

Fig. 3.4. Spectral distribution of irradiance beneath sea ice in McMurdo Sound. Snow cover reduces PAR penetration at all wavelengths; the ice algal bloom strongly attenuates light at the blue and red ends of the spectrum. Redrawn from Palmisano *et al.* (1987a),

able radiation (PAR, light within the waveband available to plants, 400–700 nm) by about 90% with peak transmission around 500 nm (Palmisano, *et al.*, 1987a). Snow cover greatly influences the availability of light for photosynthesis. For example, the irradiance beneath a quadrat of sea ice artificially shaded with 70 cm of snow was less than 3% of the PAR beneath a control quadrat with only seven centimetres of natural snow cover (Palmisano, Kottmeier & Sullivan, 1985a). PAR is also strongly attenuated by the sea-ice microalgal community which can reduce light levels to less than 0.5% of surface levels and shift the transmission peak by selectively absorbing blue and red wavebands of PAR (Fig. 3.4; further information in Palmisano *et al.*, 1987a).

## 3.3    Microbial community structure

Three broad categories of microbial community have been recognised in the sea-ice ecosystem. The bottom-ice community appears to be the most productive and includes micro-organisms which inhabit the interstices of the platelet ice layer (when present), species living in drainage channels and brine pockets at the base of the congelation ice, as well as mat- chain- and tube-forming species that are anchored to the under-ice surface and hang down into the seawater beneath. The surface-ice communities inhabit a similarly diverse range of environments including meltpools, snow infiltrated with seawater and the snow–ice–seawater mixture which accumulates in tide cracks. A third microbial group, the interior-ice community, comprises organisms which have been detected at intermediate depths within the ice sheet. This mid-depth assemblage has been commonly recorded in the Weddell Sea region.

Microalgae account for most of the community biomass and have therefore received the most attention from sea-ice investigators, but bacteria, fungi and protozoa are co-inhabitants in all three types of environment. Various terms have been proposed to describe this overall assemblage of organisms including *epontic* (meaning 'out of the sea'), *sympagic* ('with ice') and *SIMCO* ('sea ice microbial community').

### 3.3.1 Microalgal communities
#### Population structure

A wide array of microalgae, mostly pennate diatoms, have been identified in the sea-ice microbial ecosystem (Fig. 3.5). In their early studies on the ice communities of McMurdo Sound, Bunt & Wood (1963) recognised two groups of diatoms: attached forms such as *Pleurosigma, Nitzschia, Amphiprora* and *Fragilaria*, and non-attached forms such as *Biddulphia, Coscinodiscus* and *Asteromphalus*. More recent investigators in East McMurdo Sound identified the algal dominants as a chain-forming *Amphiprora* and a stellate colonial species, *Nitzschia stellata*. These species were most abundant in the loose platelet-ice underlayer. In the New Harbour region of West McMurdo Sound a tube-dwelling pennate diatom was common. This species produced mucilage-encased tubes up to 10 cm long that extended through brine channels and hung from the bottom of the hard ice into the seawater beneath (Palmisano & Sullivan, 1983b).

A similar type of hanging community has been reported off Mawson, Davis and Casey (McConville & Wetherbee, 1983). In 1981 the sea ice of this region lacked a brash-ice underlayer, but the bottom surface was coated with a mat of diatoms from which large clumps or strands hung down into the water. The strands varied from 5 to 15 cm long and were

Fig. 3.5. Pennate diatoms, such as this species of *Amphiprora* from the congelation ice in McMurdo Sound, are generally the biomass dominants in the bottom sea-ice community. Scanning electron micrograph reproduced from Palmisano & Sullivan (1983) by permission of Dr Cornelius Sullivan and Polar Biology. The individual frustules are *c*. 90 $\mu$m long and 20 $\mu$m wide.

dominated by a tube-dwelling diatom that seems to resemble the as yet unidentified (*Berkeleya*?) West McMurdo tube species (Fig. 5 in McConville & Wetherbee, 1983; cf. Fig. 7c in Palmisano & Sullivan, 1983b). Other species that were important in the suspended strands beneath the ice were *Entomoneis* (=*Amphiprora*), which grew as long ribbon-like colonies and formed an anastomosing network with the mucilage-tube producing species, and *Nitzschia frigida*. As in West McMurdo Sound, the microalgal strands were anchored by extension of the colonial diatom chains 2–3 mm into the brine channels on the undersurface of the congelation ice. A number of epiphytic pennate diatoms were also recorded in this bottom-ice community, particularly *Synedra* spp. During December and January various centric diatoms (e.g. *Stephanopyxis*) and phytoflagellates (*Chlamydomonas*) became common in this lowermost layer.

Three other types of microalgal communities were observed in the sea ice near Casey (McConville & Wetherbee, 1983). A bottom-ice assemblage colonised the brine channels in the lowermost 5 cm of the congelation ice. These channels were formed as cylindrical spaces up to 500 $\mu$m in diameter between the 1.2 mm wide, vertically-oriented ice crystals. They became densely packed with *Nitzschia frigida* and *Entomoneis* spp. in concentrations up to $3 \times 10^8$ cells $1^{-1}$. *N. frigida* formed an open lattice array within the channel that did not seem to impede the flow of brine and seawater. This congelation-ice flora was identical in structure to the bottom-ice community described in the Arctic, unlike the hanging community which does not seem to have a northern, high latitude equivalent.

The two bottom communities at Casey developed during September and persisted until ice breakup and melting in January. In this mid- to late-summer period a separate microalgal community developed near the surface of the sea ice. It occupied meltpools just beneath a 5–10 cm layer of consolidated snow (firn layer) and was primarily composed of small diatoms (*Fragilariopsis linearis, F. obliquecostata, F. ritscheri*), flagellates (*Pyramimonas, Gymnodinium, Cryptomonas cryophila, Mantoniella squamata*) and colonies of *Phaeocystis* sp. The meltpools spread rapidly, with depths up to 15 cm but were covered by a surface crust of ice that was almost completely separated from the main body of ice beneath. During early January the brine channels through the ice became interconnected and contained algal species from both the upper- and lower-ice communities. By the time of ice breakup in mid-January the meltpools contained dense, visibly coloured aggregations of microalgae, up to $2.8 \times 10^{12}$ cells $1^{-1}$.

A fourth microbial community in the Casey sea ice was possibly composed of residual populations from an autumnal-ice algal bloom. A patchy, visibly coloured band up to 35 cm thick and 20–40 cm below the surface was found in sea ice that had formed the preceding autumn. When examined in December this layer contained mostly empty diatom frustules, but also some viable microalgae that were photosynthetically active (McConville & Wetherbee, 1983).

Near Syowa Station (69°S, 39°E) two layers of micro-organisms developed as the result of separate growth phases during the year (Hoshiai, 1977). A community grew in the brine pockets in the bottom layers of hard sea ice when it was only 30 cm thick during autumn. The dominant constituents were *Nitzschia* spp., *Fragilariopsis* and various unidentified flagellates. The sea ice then thickened during winter leaving the autumnal community high in the ice profile. During spring another community grew at the base of the ice sheet, which was by then 130 cm thick. This spring assemblage was dominated by *Amphiprora, Pleurosigma, Stephanopyxis, Nitzschia* and *Navicula*. These species were not found higher in the ice profile, but by late spring the organisms of the upper community were found throughout the ice to depths just above the bottom community. This seasonal pattern of diatom growth resulted in a brown stain to the base of the autumnal ice which then moved higher in the ice column over winter. A second but darker brown colour developed in the bottom ice during the second phase of growth (Fig. 3.6).

Tide cracks may also contain rich diatom communities. Abundant, uni-algal populations of *Navicula glaciei* developed in the coastal tide cracks at Signy Island (Whitaker, 1977). These algae inhabited the snow–ice mixture infiltrated with seawater and were first detected at the end of May. Growth was slow until early September, and then with improved irradiance conditions the populations increased rapidly to a peak in early

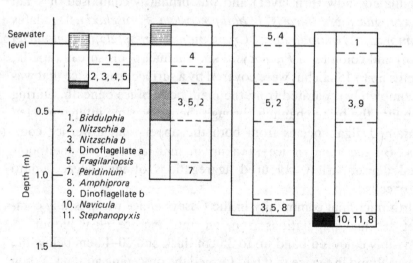

Fig. 3.6. Seasonal succession of microalgae in various strata of sea ice off Syowa Station. The left side of each ice profile shows the amount of snow cover (lined section) and light brown-stained ice (light shading) or dark brown-stained ice (dark shading). Modified from Hoshiai (1977).

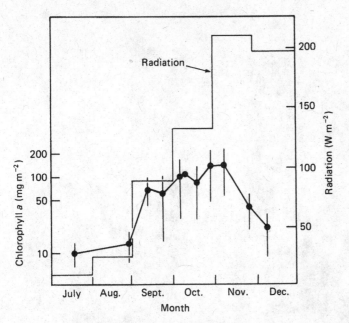

Fig. 3.7. The seasonal growth cycle of a diatom population (as measured by chlorophyll *a*) in a tide crack at Signy Island. Redrawn from Whitaker (1977); the radiation data are from Walton (1977).

November (Fig. 3.7). Beyond this period the community declined continuously until break-up of the fast ice in the first week of December. Whitaker (1977) also observed microalgae in a variety of other ice habitats, including thick (up to 5 cm) layers of a diverse (about 50 species) community dominated by *Biddulphia punctata* growing on the base of a grounded iceberg.

Pennate diatoms, but also the prymnesiophyte *Phaeocystis pouchetii* have been recorded as the numerical dominants in the pack ice of the Weddell Sea. Garrison, Buck & Silver (1983) found a close similarity in algal species composition between the Weddell Sea ice and planktonic communities, but the relative abundance of species differed considerably. Their observations suggested that phytoplankton cells were initially entrapped and concentrated during the formation of frazil ice. *In situ* growth could not account for the very high microalgal biomass (up to 50 times more concentrated than in the water) that accumulated in day-old ice. They suggest two mechanisms by which this concentrating effect could operate: nucleation and/or scavenging.

Frazil ice crystals form around suspended cells in the water column, and harvest microalgae as the crystals float to the surface. Populations in the

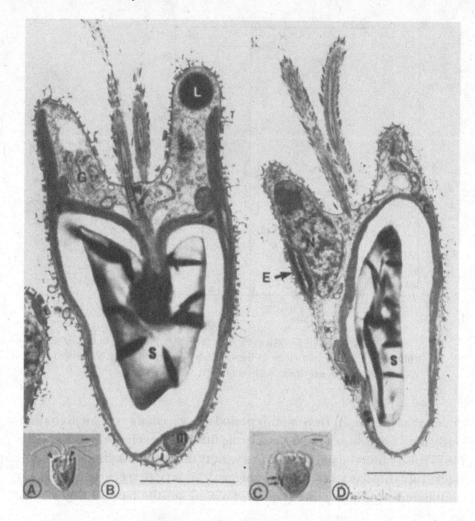

Fig. 3.8. *Pyramimonas gelidicola*, a phytoflagellate which has been recorded in bottom sea-ice communities off East Antarctica. Transmission electron micrographs (B, D) and light micrographs (A, C) reproduced from McFadden *et al.* (1982) by permission of Dr Geoff McFadden and Phycologia. The electron micrographs are of longitudinal sections at right angles to each other. Labels: C, chloroplast; E, eyespot; G, Golgi body; L, lipid body; M, mitochondrion; m, microbody; S, starch; scale bars = 3 $\mu$m.

newly forming ice in the Weddell Sea were almost identical in relative species abundance to the phytoplankton community in the water column. Garrison *et al.* (1983) speculate that this process may ensure that resting stages of pelagic algae are seasonally incorporated into the sea ice. These cells overwinter within the frazil ice and are then released into the water column during ice breakup and melt over spring and summer.

Ice nucleation and scavenging seems to be a much less important mechanism of algal accumulation in the McMurdo Sound ice. No net increase of algal populations was observed in a quadrat of ice off Ross Island experimentally darkened to less than 3% of normal irradiance levels. In an adjacent control quadrat algal cell concentrations increased 14-fold over the six week study; during which time there was a 17 cm accretion in ice thickness at both sites (Sullivan *et al.*, 1983).

Flagellates generally make only a small contribution to the total biomass of the sea-ice algal community, but on occasion they have been reported as the community dominants. *Pyramimonas gelidicola* (Fig. 3.8) was observed in high population densities in the bottom layers of sea ice from Prydz Bay (68°S, 78°E), East Antarctica (McFadden, Moestrup & Wetherbee, 1982). Related species are known to occur in the tide-crack communities in McMurdo Sound and in certain meromictic lakes on the continent (Vincent, 1981; Burch, 1987).

*Biomass structure*

The sea-ice profile of algal biomass is entirely dictated by the relative development of surface-, interior- and bottom-ice communities. In the McMurdo Sound region the former two assemblages are relatively unimportant, and most of the algal biomass is located in the platelet layer, the lowermost congelation ice or in the hanging community. In parts of the Weddell Sea neither a meltpool nor a platelet community develops and the biomass peaks midway down the profile. Off Casey, the biomass maximum has been recorded in the lower-ice region, but later in the season the peak values occur near the surface.

In 1980 the annual ice in McMurdo Sound began to form in April. By November it consisted of a 1.3–2.5 m thick layer of congelation ice with a surface layer of up to 20 cm of snow (Palmisano & Sullivan, 1983b). That year there was no platelet ice at the study sites although a 10–20 cm thick platelet layer was reported for certain other locations in the Sound (A. Palmisano, pers. comm.). Chlorophyll *a* levels rose by more than four orders of magnitude down the ice column, with a sharp maximum in the bottom 20 cm (Fig. 3.9). Concentrations averaged 656 mg m$^{-3}$ in this bottom-most stratum of congelation ice, with similarly high phaeophytin (369 mg m$^{-3}$). These pigment values were more than two orders of magnitude higher than the underlying seawater (Fig. 3.9). The ratio of phaeopigment to chlorophyll *a* averaged 0.8 in this bottom region, but was consistently greater than 3.0 at higher depths. Particulate organic carbon also rose with depth but to a lesser extent than chlorophyll *a* (Fig. 3.9), and

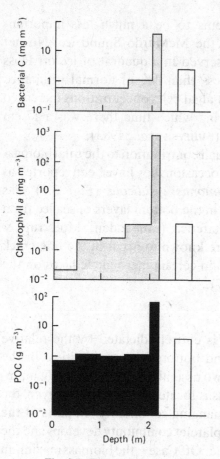

Fig. 3.9. The distribution of total microbial biomass (particulate organic carbon, POC), microalgal biomass (chlorophyll *a*) and bacterial biomass (cellular carbon) in the congelation ice at McMurdo Sound. Note the logarithmic scales. The open bars are for seawater immediately beneath the sea ice. Modified from Palmisano & Sullivan (1983) and Sullivan & Palmisano (1984).

ATP concentrations were also 2–3 orders of magnitude higher in the bottom 20 cm.

Chlorophyll *a* maxima were recorded much higher in the ice column in the Weddell Sea pack ice (Ackley *et al.*, 1979). Chlorophyll distribution followed that of salinity, with highest algal biomass and maximum salinities about midway down each core (70–100 cm below the surface). The chlorophyll *a* values recorded were all very low (maximum of 4.5 mg m$^{-3}$) by comparison with the McMurdo bottom-ice communities. Most of this chlorophyll *a* is probably derived from the water column during ice-crystal

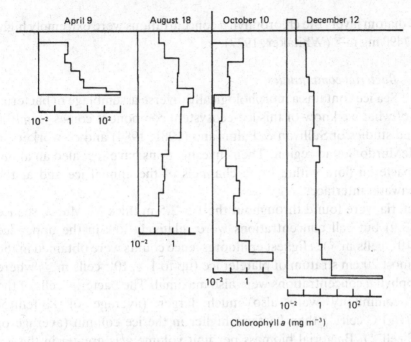

Fig. 3.10. Changes in the chlorophyll *a* profiles through the year in the sea ice off Syowa Station. The total thickness of the ice on December 12 was 1.5 m. Redrawn from Hoshiai (1977).

formation, and there is no evidence as yet that this interior community contains actively growing algal cells.

The bottom-ice community in the hard ice off Syowa Station demonstrated two biomass maxima over the year (Hoshiai, 1977). Chlorophyll *a* concentrations were initially very high in autumn (up to 829 mg m$^{-3}$), decreased over winter, and then increased during spring to extremely high values by December (>1000 mg m$^{-3}$, Fig. 3.10). The peak chlorophyll concentrations occurred in the bottom-most ice strata, but during the early period of spring growth the biomass maximum was higher in the ice profile. This upper peak contained high levels of phaeopigment and probably represented the remnant populations from the autumnal-ice algal bloom. Phaeophytin concentrations were very low in the lower strata of the ice at all times of year.

Tide-crack and meltpool algae can grow to very high population densities that visibly strain the water or snow–ice–water mixture. In the tide-crack overflow region near Signy Island the biomass levels peaked in November at 148 mg chlorophyll m$^{-2}$ and 5.5 gC m$^{-2}$ (Fig. 3.7). In the dense diatom layers the chlorophyll *a* concentrations were extremely high, up to 7490 mg m$^{-3}$ (Whitaker, 1977).

### 3.3.2    *Bacterial communities*

Sea ice contains a morphologically diverse assemblage of bacteria. Most of what we know of this ice-ecosystem component comes from the detailed studies of Sullivan & Palmisano (1981, 1984) and co-workers in the McMurdo Sound region. Their investigations have revealed an abundant bacterial flora within brine channels of the annual ice and at the ice–seawater interface.

Bacteria were found throughout the 1.5–2.5 m thick McMurdo sea ice (Fig. 3.9) but cell concentrations were relatively low in the upper ice ($4 \times 10^{11}$ cells m$^{-3}$). Highest epifluorescence counts were obtained in the lowermost 20 cm stratum of platelet ice (up to $1 \times 10^{12}$ cells m$^{-3}$) where chlorophyll *a* concentrations were also maximal. The bacterial cells of the lower community were also much larger (average of 80 femtograms (fg) C cell$^{-1}$) than bacteria higher in the ice column (average of 5 fg C cell$^{-1}$). Bacterial biomass per unit volume was greater in the ice than in the water beneath the ice. Mean bacterial-carbon concentrations in the lower 20 cm of ice were about 46 mg m$^{-3}$, but about 3 mg m$^{-3}$ in the underlying water (Fig. 3.9).

Sullivan & Palmisano (1984) recognised two main components in the sea-ice bacterial community: free-living bacteria (70% of total) and attached bacteria (30% of total). The free-living species varied considerably in cell volume and were mostly cocci or short rods. The smallest cells were 0.5 $\mu$m$^3$, but larger bacteria in the range of 3–10 $\mu$m$^3$ were more common. Some of the large cells were coated in fibrillar material resembling exocellular polymeric substances.

The attached or epibacteria were observed on the surface of live, photosynthetically active microalgae, as well as on dead diatoms and other detritus. The epibacterial flora included cocci, rod, straight and branched filament, fusiform and prosthecate species. Some of the forms seemed morphologically adapted towards the attached life style. A stalked bacterium resembling *Prosthecobacter fusiformis* (a fusiform caulobacter) possessed a well-defined holdfast, while a chain-forming rod species appeared to anchor itself in the punctae of the host diatom cell wall.

Epibacteria were also abundant on the diatom communities hanging from the ice off Casey (McConville & Wetherbee, 1983). These were mostly Gram-negative rods, although Gram-negative cocci were also common. The bacteria were attached to the mucilage sheaths of *Amphi-*

*pleura* sp. ($=$*Berkeleya*) and to the frustules of *Entomoneis* but many other diatom species (e.g. *Nitzschia frigida*) were largely devoid of epibacteria. Bacterial densities ranged from $4 \times 10^{11}$ to $2 \times 10^{13}$ cells m$^{-3}$. These cell concentrations were up to 10 times higher than in the McMurdo Sound community.

Seasonal changes in the bacterial community have been followed in the lower 20 cm of congelation ice in McMurdo Sound (Grossi, Kottmeier & Sullivan, 1984). The bacterial cell counts (by epifluorescence) increased over the spring–summer period, but net growth rates were about 0.25–0.5 of the net algal growth rates. Over an intensive six week study bacterial numbers increased six-fold, to $1.8 \times 10^{11}$ cells m$^{-2}$, whereas biomass increased 10-fold to 2.0 mg C m$^{-2}$. This disparity between cell number and biomass increments was due in part to the faster growth rates of the large (15 fg C cell$^{-1}$) epiphytic bacteria relative to the smaller (9 fg C cell$^{-1}$) unattached bacteria.

There were also qualitative changes in the epiphytic bacterial flora through the seasons. In late October the *Amphiprora* ($=$*Entomoneis*) frustules were sparsely colonised by small rods and cocci. By mid-November this community of attached bacteria had developed into a diverse assemblage of fusiform cells, filaments and prosthecate bacteria, but diatoms other than *Amphiprora* were largely uncolonised. At the decline of the ice algal bloom in January, bacteria extensively covered most of the diatom species. The epiphytic assemblage was dominated by filamentous and prosthecate bacteria, but fusiform cells, rods and cocci were also abundant. The average number of epiphytic bacteria on *Amphiprora* continued to increase over March–May, but the average biomass per cell decreased, from 25 fg C cell$^{-1}$ in January to 10 fg C in May.

The faster net growth rate of epiphytic bacteria implies a direct coupling between algal photosynthesis and bacterial production. This type of cell count data, however, poses interpretational difficulties. The bacterial communities may experience high loss rates through, for example, cell lysis or grazing by bactivorous protozoans. The measured net growth rates will underestimate true bacterial production to an unknown but potentially large and variable extent. More convincing evidence of a close relationship between algal and bacterial growth came from a perturbation experiment in which algal growth rates were artificially slowed by covering the sea ice with a thick layer of snow. The bacterial counts in the sea ice beneath this shaded quadrat showed no significant changes over the six week period while bacterial populations in the adjacently unperturbed sea ice grew rapidly (Grossi *et al.*, 1984).

### *3.3.3   Protozoan communities*

Protozoa typically rise to abundance during the late stages of development of the sea-ice microbial ecosystem. In the tide-crack overflow environment ciliates (*Chilodonella* sp., *Euplotes* spp., and a motile perit-rich) were observed feeding on diatoms late in the season (Whitaker, 1977). Similarly, numerous ciliates and flagellates appeared in the McMurdo platelet ice environment near the end of the growth season (Sullivan *et al.*, 1983).

A diverse assemblage of flagellates, amoebae and ciliates were isolated from ice samples collected in the Weddell Sea (Fenchel & Lee, 1972). Two new ciliate species were described from this study: *Spiroprorodon glacialis*, a large holotrich that fed on phytoflagellates and other ciliates, and *Euplotes antarcticus*, a non-planktonic species that fed on bacteria.

Planktonic foraminifera have been commonly found inside the sea ice. Lipps & Krebs (1974) recorded abundant populations of *Globigerina* sp., (probably *G. pachyderma*) in antarctic pack ice. The foraminifera were juveniles and in high concentration (up to $4.7 \times 10^5$ cells m$^{-3}$) suggesting that they were alive and growing within the interstices of the ice. Lipps & Krebs (1974) suggest that the ice may act as a nursery and refuge for developing foraminifera.

Choanoflagellates have been recorded under the ice in the Weddell Sea (Silver, Mitchell & Ringo, 1980). They occur in relative abundance (about $10^8$ cells m$^{-3}$) in the upper 100 m of the water column. These organisms perhaps utilise small-sized organic particles that fall out from the overlying ice-sheet communities. High concentrations of siliceous cyst-like particles were also found in this sub-ice environment in the Weddell Sea, and were initially thought to represent resting stages of the choanoflagellates. Subsequent work has shown that these minute spheroids ($<5\ \mu$m in diameter) are unrelated to the protozoa but contain a chloroplast and are actually vegetative algal cells. They appear to be widely distributed in the Southern Ocean, and are also common in the Subarctic. The chrysophyte order Parmales (referring to the 'small round shields' of silica which cover the cells) has been erected to incorporate these newly discovered algal species (Booth & Marchant, 1987). Radiolarians have also been observed in the Weddell Sea, beneath sea ice (Morley & Stepien, 1984). Like the choanoflagellates, the polycystine species were most abundant in a warm cell of Weddell Deep Water. The most common polycystine representative was *Spongotrochus glacialis*. Phaeodarian radiolarians occurred in slightly deeper habitats with maximum abundance between 100 and 200 m. The most common species in this group was *Challengeron bicorne* beneath the sea ice, but *Protocystis harstoni* at the ice edge.

### 3.3.4 *Fungal communities*

Chytrid fungi are common parasites of marine and freshwater diatoms, and at least three species have been recorded from arctic sea-ice diatoms (Horner, 1976). Chytrid fungi have been similarly observed on McMurdo Sound diatoms (R. L. Moe, unpublished observation). Saprophytic fungi of the antarctic sea ice have been unexplored.

### 3.4 Microbial processes

### 3.4.1 *Photosynthesis*

Many authors have suggested the importance of sea-ice algal production to the overall carbon economy of the Southern Ocean. However, photosynthetic data are sparse and are mostly available for the productive McMurdo Sound community which may not be representative of the entire sea-ice zone.

In a detailed series of photosynthetic experiments the McMurdo Sound brash-ice algae were found to have a very low compensation point (well below 1% of surface irradiance) and also a low productivity per unit biomass (Bunt, 1968; Bunt & Lee, 1970). Over a season of sampling the mean maximum assimilation number (mg C (mg chl $a$)$^{-1}$ h$^{-1}$) was 0.24, and the mean minimum was 0.056 at light intensities that saturated photosynthesis. However, with light-adapted cultures Bunt (1968) recorded assimilation numbers as high as 1.2.

Algal production has been recorded in the ice off Casey (McConville & Wetherbee, 1983), although rates were expressed per unit volume and are therefore difficult to assess on an areal basis. Photosynthetic carbon uptake by the bottom community ranged from a maximum of 81 $\mu$g C l$^{-1}$ h$^{-1}$ during November to 2.8 $\mu$g C l$^{-1}$ h$^{-1}$ in mid-January. Much higher rates of production (up to 363 $\mu$g C l$^{-1}$ h$^{-1}$) were recorded in the surface meltpool community containing small diatoms, flagellates and *Phaeocystis*. The interior coloured band (see above) also had measurable photosynthesis, $c$.5 $\mu$g C l$^{-1}$ h$^{-1}$. All of these production measurements were performed at a single irradiance (24 $\mu$E m$^{-2}$ s$^{-1}$) and at about $-1$ °C.

Microalgal population growth was followed by direct cell counts in the McMurdo Sound brash-ice community during 1962–63 (Bunt, 1963). Growth rates were highest in mid-December (0.19 natural log units $\cdot$ d$^{-1}$) and were comparable with phytoplankton growth rates in temperate oceans. However, with uni-algal cultures Bunt (1968) obtained growth rates at 7 °C of up to 1.2 d$^{-1}$ after adaptation to bright light.

Several investigators have estimated annual production rates on the basis of maximum algal standing crop. This type of approach clearly underestimates total production for it does not include dissolved organic carbon produced and released by the diatom cells, nor any other loss components such as natural mortality, grazing, sloughing or parasitism. Bunt & Lee (1970) found an extreme patchiness in the brash-ice community, and from standing crops estimated a net annual production in a shaded station of 1.08 g C m$^{-2}$, and 0.52 g C m$^{-2}$ at a sunny station. Palmisano & Sullivan (1983b) recorded an average chlorophyll $a$ level in McMurdo Sound of 131 mg m$^{-2}$ and from an average chlorophyll : carbon ratio of about 30:1 they derived a minimum annual production figure of 4.1 g C m$^{-2}$. Later experimental work by this research group using *in situ* photosynthetic assays has shown that the actual net production rate per annum is probably 10 times higher than this earlier estimate (Grossi *et al.*, 1987).

The Signy Island tide-crack community annual production has been similarly estimated from standing crop at 1–2 g C m$^{-2}$. This was very small by comparison with annual phytoplankton production in the region (80–130 g C m$^{-2}$, Horne, Fogg & Eagle, 1969). The overall photosynthetic contribution from this surface-ice community was therefore considered a negligible part of total production in the South Orkney Islands (Whitaker, 1977), but perhaps unjustifiably given the comparison of methods in McMurdo Sound.

### 3.4.2    *Photoadaptation and nutrient deficiency*

Early work in McMurdo Sound emphasised the highly shade-adapted characteristics of the platelet-ice microalgae (Bunt, 1963, 1968). Diatoms sampled from the platelet layer were capable of 20% of their maximum photosynthetic rates at the dim ambient light intensities (8 foot-candles, about 2 $\mu$E m$^{-2}$ s$^{-1}$) in the zone of collection. The algae were saturated at only 100 foot-candles (about 22 $\mu$E m$^{-2}$ s$^{-1}$) and were photoinhibited at 1100 foot-candles (about 230 $\mu$E m$^{-2}$ s$^{-1}$).

More detailed photosynthesis–irradiance investigations in McMurdo Sound have confirmed the extreme 'shade' characteristics of the platelet community. This community in December 1983 had a low $I_k$ (irradiance of the onset of saturation; average of 7 $\mu$E m$^{-2}$ s$^{-1}$), low maximum assimilation number (photosynthetic rate in terms of carbon fixed per unit weight of chlorophyll $a$; <0.1 mg C (mg chl$a$)$^{-1}$ h$^{-1}$) and was photoinhibited (bright light depression of photosynthesis) at irradiances above 25 $\mu$E m$^{-2}$ s$^{-1}$. All of these features show that the community was adapted to very low ambient light levels. When the sample site was covered by 4 cm of snow the assimilation dropped further, to 60% of initial values over 2 days (Palmisano, Soo Hoo & Sullivan, 1985b).

The photosynthetic properties of the platelet community differ considerably from the algae at higher depths in the ice column, and Palmisano & Sullivan (1985b) conclude that the physiological ecology of the communities of each stratum should be considered separately. A surface tide-crack community dominated by *Phaeocystis pouchetii* had a high assimilation number (3.94 mg C (mg chl$a$)$^{-1}$ h$^{-1}$) and the end-products of photosynthesis were primarily protein and polysaccharide. By contrast the platelet-ice community had a low assimilation number (0.15 mg C (mg chl$a$)$^{-1}$ h$^{-1}$) and the percentage of $^{14}$C–CO$_2$ incorporated into small molecular weight metabolites was about twice that for the surface community. A third community, in the bottom congelation ice, had assimilation numbers comparable with the platelet community but the incorporation of carbon-14 into protein was only one-quarter to one-fifth of that for the surface and platelet communities. Percent incorporation into polysaccharides was 2–3 times higher for this congelation community than in the surface and platelet zone. The flow of carbon into polysaccharide at the expense of protein is characteristic of algal populations experiencing a cellular shortage of nitrogen. The labelling results from McMurdo Sound imply that the algae growing in the minute brine pockets of congelation ice may exhaust their available nutrients and become nitrogen deficient (Palmisano & Sullivan, 1985b).

Similar indications of potential nutrient deficiency have been obtained in the annual congelation ice off Davis and Mawson (McConville, Mitchell & Weatherbee, 1985). In these algal communities during December a major proportion of the $^{14}$C-label (mean of 53%) was incorporated into a $(1 \rightarrow 3)$-glucan. This reserve glucan accounted for 93% of the cellular carbohydrate of the ice algae, and its preferential synthesis relative to protein suggests nutrient limitation.

### 3.4.3 *Heterotrophic production*

The first measurements of bacterial heterotrophy in the sea ice were performed using various radio-labelled organic substrates. In the bottom-ice communities off Casey the turnover times for $^3$H-labelled amino acids ranged from 15 to 154 h (McConville & Wetherbee, 1983). The shortest turnover times occurred in the communities with the highest

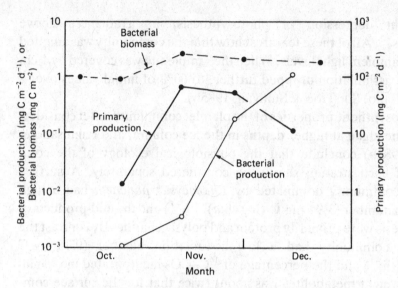

Fig. 3.11. Bacterial and microalgal production rates in the bottom 5 cm of congelation ice in McMurdo Sound. In this layer the net increase in bacterial biomass was small relative to production rates, suggesting that the cell populations were controlled by bactivorous protozoa. Numerous ciliates and amoebae were observed in this ice in the later stages of the algal bloom. Modified from Kottmeier *et al.* (1987).

photosynthetic rates. Fractionation of the particulate material showed that most of this label was incorporated into the $>2 \mu$m size fractions, with 1–6% in the 0.2–2 $\mu$m fraction. These data suggest that either the attached bacteria were the dominant microheterotrophs in this sea-ice community and the free-living bacteria were relatively unimportant, or that the microalgae were capable of organic nutrient uptake.

More recently, bacterial growth rates in the McMurdo Sound sea ice have been directly measured by epifluorescence cell counts and [3]H-thymidine assays (Kottmeier, Grossi & Sullivan, 1987). The two approaches gave similar estimates of growth rates in the platelet ice, but not in the congelation ice where a major loss process, probably grazing by protozoa seemed to be operating (Fig. 3.11). In both communities bacterial growth rates increased through the course of the diatom bloom to a maximum of about 0.2 $d^{-1}$. Although bacterial growth was closely coupled to the development of the phototrophic community, the final bacterial biomass was less than 1% of the microalgal biomass and bacterial carbon production was at most only 9% of photosynthetic carbon fixation.

### 3.4.4 *Winter survival*

Algal heterotrophy has been examined by several investigators as a potential survival mechanism that would allow the sea-ice algae to sustain themselves during winter darkness. Bunt & Lee (1972) found no significant growth of several sea-ice algal isolates on acetate, pyruvate, glycolate, succinate, citrate, lactate or glycine after 10 months of sub-compensatory light intensities. However, in a later study of three McMurdo sea-ice diatoms there was a dramatic increase in heterotrophic potential (measured by glucose uptake from a 1 $\mu$M solution) as the cells were taken across a simulated summer–winter transition in irradiance, temperature and salinity regime. A clone of *Nitzschia cylindricus* increased its uptake capacity by 60-fold during the transition (Palmisano & Sullivan, 1982). The correct preincubation conditions may therefore need to be chosen very carefully to adequately test organic nutrient uptake by the sea-ice flora. However, the source of the algal inoculum for sea-ice communities, and its *in situ* mode of carbon and energy supply during the overwintering period still remains uncertain (see Chapter 10).

### 3.5  Trophic structure

Sea ice provides an important habitat and food source for organisms at several trophic levels. Diatom photosynthesis drives the ecosystem by producing organic carbon, which is then available to heterotrophic micro-organisms such as bacteria, fungi, foraminifera (e.g. *Globigerina*; Lipps & Krebs, 1974), ciliates and flagellates, and herbivorous animals such as harpacticoid and calanoid copepods (e.g. *Harpacticus furcifer, Idomene antarctica*; Bradford, 1978), amphipods (e.g. *Pontogeneia antarctica*; Richardson & Whitaker, 1979) and euphausiids (see El-Sayed, 1971). Various carnivores feed on the small animals, notably fish fingerlings (e.g. *Trematomus borchgrevinki*, Gruzov, Propp & Pushkin, 1967) and amphipods (e.g. *P. antarctica*, Richardson & Whitaker, 1979). The euphausiids are eaten by the larger animals of the Southern Ocean – penguins, petrels, seals, whales and adult fish. The fish provide food to seals and certain whale species.

Although these trophic pathways from the sea ice to the open ocean undoubtedly exist, their efficiency of carbon and energy transfer is completely unknown. The idea of a direct coupling between the ice and planktonic communities mediated by a sea-ice fauna may turn out to be a misleading simplification. The sea-ice algal bloom is highly seasonal and would require a very fast growth response by the herbivores to be substantially transferred into higher trophic levels. Only a limited portion of the ice community is accessible to large herbivores. The calanoid copepod *Paralabidocera antarctica*, for example, has been found to swarm near the lower surface of fast ice. Its absence from the water column at night, and the observed high concentration of copepod fecal pellets packed with pennate diatoms has provided strong circumstantial evidence that this species enters the loose ice layer each night to feed on the lowermost microalgal community (Tanimura *et al.*, 1984). However, it is unlikely that it penetrates very far into the ice. Higher in the ice column the phototrophs are isolated from the seawater beneath and are completely protected from large grazers.

The critical period that exposes the entire ice community to herbivores and detritivores is during the melt each year. This melting process occurs rapidly, and it radically alters the physical, chemical and biological properties of the ocean in the vicinity of the retreating ice edge. The dynamics of this marginal ice-edge environment is the subject of the next chapter.

# 4

# The marginal-ice zone

## 4.1 Introduction

The first compilations of photosynthetic data from the Southern Ocean produced estimates of the overall productivity of the region that seemed puzzlingly low. Antarctic whaling expeditions, and the oceanographic cruises earlier this century had brought back vivid accounts of prolific krill, seals, whales and birdlife, and it was inferred that similarly high levels of primary production would be required to support these large standing stocks of marine animals. This traditional but unverified concept of a highly productive Southern Ocean had persisted for many decades, and had been reinforced by observations of dense phytoplankton blooms at various coastal sites around Antarctica. However, when the direct carbon-14 assays of algal photosynthesis were compiled from many parts of the region it became apparent that throughout most of the Southern Ocean primary production is low relative to other oceanic upwelling areas ($<0.15$ g C m$^{-2}$ d$^{-1}$), but comparable with moderately oligotrophic waters elsewhere (Hewes, Holm-Hansen & Sakshaug, 1985).

This inconsistency between predicted and measured production rates further widened when oceanographers carefully examined the extent of nutrient depletion in the Weddell Sea (Jennings, Gordon & Nelson, 1984). The summer decrease of nitrate, phosphate and silicate in these waters implied that the average seasonal primary productivity of the Weddell region was 0.2–0.4 g C m$^{-2}$ d$^{-1}$. These estimates were 1.5–5 times above the values previously obtained in the Weddell Sea by C-14 photosynthetic assays, seemingly outside the limits of error of this technique.

More recent work has begun to reconcile some of these major inconsistencies. Intense research by several nations has begun to build up a picture of the most dynamic microbial habitat in the Southern Ocean; the marginal zone at the edge of the receding ice pack. This region has distinct physical and chemical properties, and experiences brief periods of intense biological activity that probably contribute a major part of the total photosynthetic production in Antarctica.

When the sea ice melts each year a marginal-ice zone sweeps across the Southern Ocean. Most of the Weddell Sea, for example, experiences an ice-edge environment for some period during spring (Jennings *et al.*, 1984). This interface between seawater and melting ice is a frontal region of complex physical exchanges. Strong vertical gradients of temperature and salinity are established which enhance the availability of light for algal photosynthesis by decreasing the extent of mixing of cells below the euphotic zone. These short-lived conditions of increased water column stability appear to favour a brief but intense phytoplankton bloom which in turn stimulates the activity of other microbial groups and trophic levels.

The biological and physical processes in this frontal zone, and the way in which they are coupled, still remain poorly understood. Much of what is currently known has been derived from a small number of multidisciplinary cruises. Oceanographic data from the US–Soviet Weddell Polynyna Expedition in the antarctic spring allowed the first attempts to compare seasonal nutrient depletion with previous direct assays of primary productivity (Jennings *et al.*, 1984). A large collaborative venture in 1983 involved 41 scientists on board two ships in the southern Scotia and northwestern Weddell Seas. This and subsequent cruises within the AMERIEZ programme (Antarctic Marine Ecosystem Research at the Ice Edge Zone) have provided detailed descriptions of microbial as well as higher trophic level characteristics of this habitat (Ainley & Sullivan, 1984), although by necessity at a restricted set of locations and times of the year. The Ross Sea ammonium flux experiment in 1982 focussed on the $NH_4$ content of the ice-shelf water that flows northward from beneath the Ross Ice Shelf. This barrier zone is another ice-edge region where extensive algal blooms have been reported, but its physical and chemical properties differ from the sea-ice margin. A number of smaller-scale cruises have concentrated on specific aspects of the structure and functioning of these highly seasonal microbial ecosystems.

## 4.2    The environment

The marginal-ice zone is a dynamic frontal system where air, ice and sea interact. It is not, however, a sharply delimited line between the ice and the open sea, but rather a wide transitional zone with large spatial (especially north–south) and temporal differences in its environmental properties. The pack ice extends at least 100 km outward from the edge of

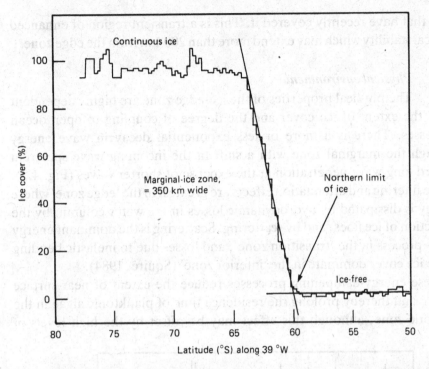

Fig. 4.1. North–south distribution of pack ice in the marginal-ice zone in the Weddell Sea, November 21 1983. This high level of spatial detail has been obtained from satellite microwave observations. Redrawn from Comiso & Sullivan (1986).

the continuous sea ice (Fig. 4.1), and surface conditions in this region vary from high concentrations of ice floes to wide areas of open water. Studies from the Arctic Ocean have separated this ice-containing region into three zones (Squire, 1984):

(i) the edge zone which lies within about 5 km of the open water. In this zone there is sufficient wave energy to fracture and break up the sea ice into small floes and ice cakes.

(ii) the transition zone which extends from 5 to 30 km or more. Here the waves fracture the ice but do not have sufficient energy to cause floe impaction and collisions. The ice floes in this region are much larger, typically of the order of 50–100 m across.

(iii) the interior zone where the ice is almost continuous and is subject only to long period swells.

A fourth zone may be identified that is of special microbial interest: the region beyond the pack ice that is influenced by meltwater from the ice floes that have recently covered it. This is a transient region of enhanced vertical stability which may extend more than 200 km from the edge zone.

### 4.2.1    *Physical environment*

The physical properties of the ice-edge zone are highly dependent upon the extent of ice cover and the degree of coupling to open ocean processes. There is a more or less exponential decay in wave energy through the marginal zone with a shift in the incoming wave spectrum toward long wave penetration at the expense of shorter waves (Fig. 4.2). These filtering and attenuation effects are greatest in the 'edge zone' where energy is dissipated by hydrodynamic losses in the water column, by the impaction of ice floes, and by scattering. Scattering is the dominant energy decay process in the 'transition zone', and losses due to inelastic bending of the ice cover dominate in the 'interior zone' (Squire, 1984).

These energy-dampening processes reduce the extent of near-surface mixing and thereby prolong the residence time of planktonic algae in the euphotic zone, although this effect may be offset by the high levels of

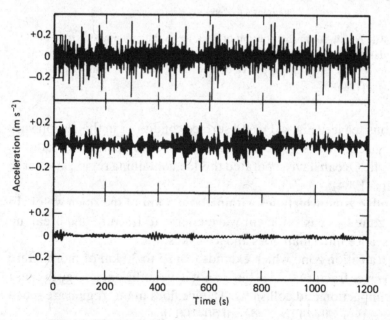

Fig. 4.2. Time series showing the energy decay of ocean waves travelling through the marginal-ice edge zone, Bering Sea. The top record is for the ocean at the outer edge of the pack-ice zone, the bottom record is for 60 km inside the pack ice and the middle record is for an intermediate station. Redrawn from Squire (1984).

shading created by the broken ice cover. Additionally, as the ice floes warm they release less saline meltwater which floats to the surface and further promotes vertical stability. This meltwater influence on the density structure of the water column may persist for days to weeks and may therefore be encountered well into open water beyond the receding ice edge. In the Ross Sea a layer of low salinity, low density water was found to extend 250 km beyond the edge of the continuous ice cover (Smith & Nelson, 1985), and similar observations have been made in the Weddell Sea (Nelson, Gordon & Smith, 1984).

The thicker icebergs derived from calving ice sheets feeding into the Southern Ocean may have an opposite effect on water column stability by generating convection. For example, an iceberg at 65 °S, 81 °E that was 280 m long with a freeboard of 25 m and an estimated draft of 200 m appeared to be melting at all depths. The water samples from the immediate vicinity of the iceberg, however, were saltier than would be expected from a mixture of meltwater and seawater and indicated that convective upwelling was bringing more saline water to the surface (Allison, Kerry & Wright, 1985). Vertical convection alongside icebergs has also been supported by laboratory and theoretical studies and identified as a potential mechanism of localised upwelling of nutrient-rich deep water (Neshyba, 1977).

### 4.2.2 Nutrients

Measurements in the Weddell Sea during the first AMERIEZ cruise showed that nutrient levels were high at all stations: about 23 mmol m$^{-3}$ nitrate, 1.4 mmol m$^{-3}$ phosphate and 75 mmol m$^{-3}$ silicic acid. These concentrations were only slightly reduced in the region of maximum algal biomass. Some reduction in $NO_3$–N has also been observed in a marginal-ice zone in the Ross Sea (Nelson & Smith (1986); see section 4.4). In the north polar region wind-driven upwelling of nutrients has been implicated as a possible mechanism promoting the ice-edge blooms (e.g. Niebauer & Alexander, 1985). However, this effect cannot be important for the Weddell Sea bloom given the persistent surplus of nutrients.

Localised depletion of ammonium appears to occur in some of the ice-edge environments. Oceanographic surveys in the 1970s encountered algal blooms in the surface waters next to the Ross Ice Shelf. At this seaward edge of the ice shelf 'the Barrier' rises to 20 m above sea level and extends 120–140 m below the sea surface. Near-surface concentrations of $NH_4$ were less than 0.2 mmol m$^{-3}$ in the region of the bloom. Much higher concentrations of ammonium, up to 1.4 mmol m$^{-3}$, were located at the base of the ice sheet as a discrete maximum, suggesting the advection of NH$_4$–rich water from beneath the Ross Ice Shelf (El-Sayed, 1984).

There is a complex pattern of water exchange beneath the Ross Ice Shelf (Fig. 4.3; Table 4.1). The highest ammonium concentrations are believed to be associated with *shallow ice shelf water* (SISW) that is formed by a localised *warm core* (WMCO) which causes substantial melting of the undersurface of the ice shelf (Jacobs, Fairbanks & Horibe, 1985). A much smaller amount of melting is induced by *high salinity shelf water* (HSSW) but the resultant current remains very deep and ultimately gives rise to *antarctic bottom water* (ABW, Fig. 4.3) which plays an important role in the global oceanic heat balance (see Chapter 5).

The detailed Ross Sea Ammonium Flux Experiment in 1982 more clearly defined the dynamics of NH$_4$–N near the Barrier. A tongue of ammonium-rich water extended at least 30 km to the north of the Ross Ice Shelf with concentrations in excess of 0.8 mmol m$^{-3}$ near the lower lip of the ice sheet. This water could potentially be transferred into the mixed layer by internal wave activity or by the passage of eddies where it would help sustain the large populations of phytoplankton that occur in the vicinity of the ice-shelf edge in late summer (Biggs, Amos & Holm-Han-

Table 4.1. *Major water masses in the vicinity of the Ross Ice Shelf as shown in Fig. 4.3*

Depth refers to the average depth of sampling for the temperature and salinity data presented. LSSW is not included in Fig. 4.3, but is an important water mass on the eastern side of the ice shelf where it replaces HSSW.

| Abbreviation | Water mass | Depth (m) | Potential temperature (°C) | Salinity (‰) |
|---|---|---|---|---|
| ASW | Antarctic surface water | 7 | −0.96 | 34.12 |
| ABW | Antarctic bottom water | 1583 | −0.16 | 34.65 |
| CDW | Circumpolar deep water | 622 | 1.17 | 34.70 |
| DISW | Deep ice shelf water | 382 | −2.03 | 34.68 |
| HSSW | High salinity shelf water | 556 | −1.91 | 34.84 |
| LSSW | Low salinity shelf water | 258 | −1.59 | 34.53 |
| SISW | Shallow ice shelf water | 119 | −2.04 | 34.36 |
| WMCO | Warm core | 312 | −0.84 | 34.54 |

Compiled from Jacobs *et al.*, 1985.

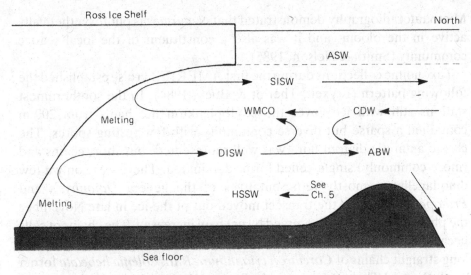

Fig. 4.3. A schematic model of water circulation beneath the Ross Ice Shelf based on the information presented in Jacobs *et al.* (1985). There is also considerable east–west variation in the distribution of the different water masses, which are described in Table 4.1. Not drawn to scale.

sen, 1985). It has also been suggested that these high concentrations of $NH_4$ derived from the melting of the ice (see Chapter 2) may support a population of planktonic nitrifying bacteria growing under the Ross Ice Shelf (Horrigan, 1981).

Nutrient concentrations across the Weddell Sea abruptly decreased at latitudes greater than 72 °S, and inversely corresponded with chlorophyll *a* levels (El-Sayed & Taguchi, 1981). Nitrate dropped from 25–30 to <10 mmol m$^{-3}$, phosphate dropped from >2.0 to <1.0 mmol m$^{-3}$, and silicate declined from 50 to <30 mmol m$^{-3}$, while chlorophyll *a* levels increased from <10 mg m$^{-3}$ north of 71 °S, to >20 mg m$^{-3}$ south of this latitude. The remaining nutrient concentrations were still relatively high, however, by comparison with oligotrophic oceans at warmer latitudes.

## 4.3    Microbial community structure
### 4.3.1    *Microalgal communities*
#### *Population structure*
Many of the ice-edge phytoplankton are also important constituents of the sea-ice community, suggesting that the sea ice provides the inoculum for the marginal-ice zone phytoplankton bloom. In the marginal-ice edge community off southern Victoria Land the phytoplankton dominant was the diatom *Nitzschia curta* which accounted for up to 85% of the total cell count. *Nitzschia closterium* was an important subdominant.

Microautoradiography demonstrated that *N. curta* was photosynthetically active in the bloom, and it was also a constituent of the local sea-ice community (Smith & Nelson, 1985).

Taxonomic collections during the first AMERIEZ cruise established the following pattern (Fryxell, Theriot & Buck, 1984). In the southernmost stations still with ice cover the phytoplankton net hauls from 200 m contained a sparse but diverse community with few resting spores. The classic antarctic diatom flora was well represented, but short chains and (more commonly) single-celled forms dominated. There were only a few dinoflagellates, mostly representatives of the genera *Dinophysis* and *Protoperidinium*. As the transect moved out of the ice in late November the phytoplankton levels sampled by net haul increased. The chains of cells became much longer including spiral chains of *Eucampia antarctica*, and long straight chains of *Corethron criophilum*, *Rhizosolenia hebetata* forma *semispina* and *Thalassiosira tumida*. Beyond the ice edge in the core region of the bloom, the dominants were two gelatinous colonial species, the diatom *Thalassiosira gravida* (Fig. 4.4) with up to several hundred frustules

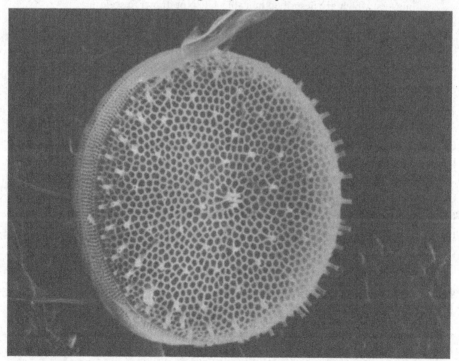

Fig. 4.4. *Thalassiosira gravida* from the Southern Ocean. This diatom has been recorded as a dominant phytoplankton species in an ice-edge bloom in the Weddell Sea. Scanning electron micrograph by Eva Braaten, courtesy of Dr Grethe R. Hasle, Department of Biology, University of Oslo.

per colony and the prymnesiophyte *Phaeocystis pouchetii*. Inshore or ice-edge communities dominated by *Phaeocystis* have been observed at various other sites in Antarctica. For example, extensive blooms of this species commonly occur in the Ross Sea after the sea ice breaks up. Chlorophyll *a* levels reach up to 14 mg m$^{-3}$ (Palmisano *et al.*, 1986).

*Biomass structure*

A wide band of high algal biomass levels has been recorded in the marginal-ice zone at many sites around Antarctica. These have included the ice edges southeast of New Zealand (Nelson & Gordon, 1982), in the Ross Sea off southern Victoria Land (Smith & Nelson, 1985), in the Weddell Sea (Nelson *et al.*, 1984) and in the sea off Wilkes Land in East Antarctica (Garrison & Van Scoy, 1985). From estimates of the seasonal depletion of nutrients in the Weddell Sea, Jennings *et al.* (1984) calculated that there should be an ice-edge bloom of sufficient magnitude to double the previous annual estimates of productivity. During the AMERIEZ cruise a phytoplankton bloom of approximately this magnitude was encountered. The core of the production maximum contained more than 5 mg chl*a* m$^{-3}$ and was located about 100 km to the north of the 'ice-edge' as defined by the maximum north–south gradient in ice cover. The high biomass region of >1 mg chl*a* m$^{-3}$ extended north–south over about 250 km and was associated with a pronounced depletion of nitrate (Fig. 4.5). Analysis of the salinity data confirmed that the recorded nutrient depletion could not simply be ascribed to dilution by water from the melting sea ice. The chlorophyll accumulation appeared to be the result of active phytoplankton growth in the water column rather than a release of algal cells from the sea-ice community.

The ice-edge microbiota in the Ross Sea region appeared to have an unusual elemental composition (Smith & Nelson, 1985). Chlorophyll *a* levels off the coast of southern Victoria Land rose to *c*. 5 mg m$^{-3}$, comparable with productive coastal regions elsewhere in the world. However, the particulate carbon levels were disproportionately higher and more typical of productive systems such as the Peruvian upwelling. Concentrations of biogenic silica (silica associated with diatoms and silico-flagellates) were the highest recorded anywhere in the world. Laboratory cultures of microalgae have carbon : chlorophyll *a* ratios in the range of 25–40 (by weight), but the mean C : chl*a* ratio for the ice-edge euphotic zone was 118.2. This ratio is known to increase under conditions of nutrient limitation or low temperatures. Since nutrient levels were high but temperatures low, possibly the latter effect was manifested in this

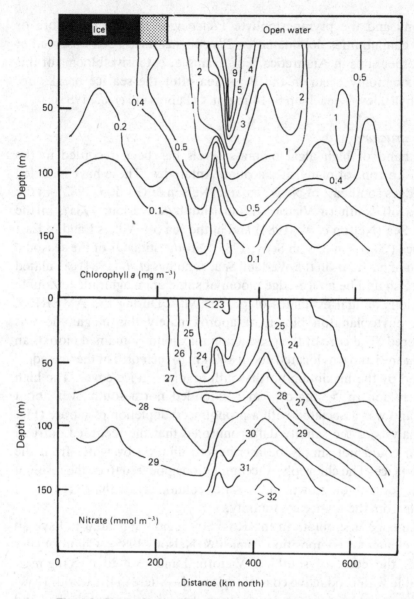

Fig. 4.5. Chlorophyll *a* accumulation and nitrate depletion in a marginal-ice zone in the Weddell Sea, November 1983. Redrawn from Nelson *et al.*, 1984.

marginal ice-edge environment. Oceanic phytoplankton generally contain biogenic silica and organic carbon in the atomic ratio 0.13, but for the ice-zone phytoplankton this ratio averaged 0.62. Antarctic diatoms appear to be heavily silicified reflecting the silicate-replete environment in which they grow. These unusual elemental ratios, however, were not observed in the Weddell Sea.

### 4.3.2 Bacteria

Detailed studies of the ice-edge bacterial community began with AMERIEZ in the southern Scotia and northwestern Weddell Seas (Miller *et al.*, 1984). Integral bacteria cell concentrations to a depth of 150 m ranged from $7 \times 10^{12}$ cells $m^{-2}$ (*c.* 100 mg C $m^{-2}$) at the southernmost stations, about 120 nautical miles into the pack ice, to $33 \times 10^{12}$ cells $m^{-2}$ (*c.* 400 mg C $m^{-2}$) in the northern stations up to 160 nautical miles beyond the ice-edge (see Section 4.4.3).

### 4.3.3 Protozoa

Choanoflagellates appeared in the net hauls in the northernmost stations during the AMERIEZ transects. Populations of protozoans were less developed in the water column than they were in the sea ice (Garrison, Buck & Silver, 1984). Population biomass was generally low under the ice ($<10^6$ organisms $m^{-3}$, about 1 mg C $m^{-3}$) but increased dramatically at the ice edge to $>10^7$ organisms $m^{-3}$ ($>10$ mg C $m^{-3}$). This distribution followed the pattern for algal and bacterial populations across the marginal ice zone (see Fig. 4.7). The dominant forms throughout were naked ciliates. Tintinnids only occurred in measurable numbers near the ice edge and heterotrophic flagellates were also more abundant there. The protozoan abundance in this marginal zone was comparable with temperate regions.

## 4.4 Microbial processes

### 4.4.1 Ice-edge photosynthesis

Jennings *et al.* (1984) calculated that a brief (10–15 d) ice-edge bloom with photosynthetic rates of 1 g C $m^{-2}$ $d^{-1}$, followed by sustained production at much lower rates (*c.* 0.1 g C $m^{-2}$ $d^{-1}$) would result in an average seasonal productivity consistent with the measured rates of spring–summer nutrient depletion. Assays of photosynthetic carbon fixation in the marginal ice-edge zones have confirmed the existence of brief episodes of algal production approaching these high upper estimates.

In late summer 1977 the algal production of the Weddell Sea was approximately four times higher at the southern stations toward the ice edge than in the northern and central regions (El-Sayed & Taguchi, 1981). Primary production values $<0.5$ mg C $m^{-3}$ $d^{-1}$ were characteristic of the northern stations, with an average rate, integrated over time and depth, of 0.104 g C $m^{-2}$ $d^{-1}$. The southern stations averaged 0.41 g C $m^{-2}$ $d^{-1}$, with highest productivities up to 0.68 g C $m^{-2}$ $d^{-1}$, off the Filchner Ice Shelf.

Several hypotheses have been advanced to account for the elevated production rates in the marginal-ice zone:

1. *The marginal zone is a region of upwelling and enhanced nutrient supply.*

   Ice-edge upwelling is supported by observational data as well as numerical models, but the effects on phytoplankton nutrition in Antarctica seem to be much less important than in the Arctic. The Southern Ocean contains extremely high concentrations of inorganic nitrogen and phosphorus, and algal growth rates are unlikely to respond to further increases in N or P supply. By contrast, nutrient levels are often low in the surface water of the arctic seas, and phytoplankton may be more responsive to upwelled nutrients there.

2. *The melting sea ice liberates its large resident community of micro-algae, thereby providing a sudden increase in algal concentration.*

   Sea ice may contain rich levels of chlorophyll *a* and this sizeable input of cells could at the very least provide an important inoculum for subsequent growth. Similarities between the planktonic and ice algal flora provide supporting evidence of this effect, but its magnitude of importance remains unclear.

3. *Lower salinity meltwater from the receding ice edge produces a stable density layering that retains the phytoplankton community under improved light conditions.*

   Low mean irradiance is thought to be an important constraint on phytoplankton growth in the Southern Ocean where the water column is generally well mixed to depths below the euphotic zone. Reduced mixing rates would increase the average light availability for the phytoplankton assemblage and thereby enhance net photosynthesis for the water column.

4. *Wind and wave energy is dissipated by the presence of ice which enhances the vertical stability of the water column, and thereby favours algal growth, as in 3.*

A series of detailed measurements off the receding ice-edge near south Victoria Land provided an opportunity to examine these competing hypotheses (Smith & Nelson, 1985). The phytoplankton bloom in this marginal zone appeared to be physically coupled to the region of meltwater influence. Algal population densities were maximal in the meltwater lens (Fig. 4.6), and were dissipated by vertical and horizontal mixing processes at the outside of this lens. The spatial coherence between algal biomass distribution and the stable layer of meltwater led Smith & Nelson (1985) to

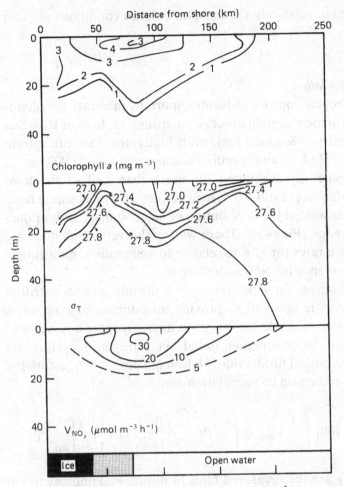

Fig. 4.6. Chlorophyll *a* distribution (in mg m$^{-3}$), water column stability (distribution of seawater density expressed as $\sigma_T$, defined as 1000 (density $-$ 1)) and nitrate uptake rates ($V_{NO_3}$ in $\mu$mol m$^{-3}$ h$^{-1}$) in a marginal-ice zone in the Ross Sea, January/February 1983. Redrawn from Nelson & Smith (1986).

favour the third hypothesis above. They further suggested that the bloom was initiated by the release of a seed population of *Nitzschia curta* from the decomposing sea ice. This pennate diatom was the dominant bloom alga, as well as a prominent member of the local sea-ice community. The fourth hypothesised mechanism (ice-stabilised water column) may also have contributed to the bloom but it appears to have much less influence than mechanism three because the region of high algal biomass extended well beyond the zone of ice cover. High biomass levels and production have also been recorded in relatively stable water column conditions at other ice-edge sites.

### 4.4.2    *Nitrate assimilation*

Nitrate-nitrogen appears to be an important substrate for phytoplankton nutrition in the marginal ice-edge environment. In their Ross Sea study Nelson & Smith (1986) found relatively high rates of specific nitrate uptake (mean of 0.0020 $h^{-1}$) and specific ammonium uptake (0.0033 $h^{-1}$), however nitrate generally accounted for more than half of the total inorganic-N assimilation (*f* ratio >0.5, see Chapter 5). The absolute rates of nitrate uptake paralleled the distribution of algal biomass and productivity near the ice edge (Fig. 4.6). There was a relatively small decrease with depth in uptake rates for nitrate relative to ammonium, indicating a stronger light dependence for $NH_4$ assimilation.

The fractional contribution of nitrate towards the nitrogenous nutrition of the phytoplankton is believed to provide an estimate of how much primary production is available for export to higher trophic levels ('new' production) relative to production based on nitrogen recycling via ammonium ('regenerated') production. Nelson & Smith (1986) estimated the extent of new production by the relationship:

$$\% \ new = 100 \left( \int_{100}^{0.1} V_{NO_3} \cdot \left[ \int_{100}^{0.1} POC \bigg/ \int_{100}^{0.1} PON \right] \right) \bigg/ \int_{100}^{0.1} V_C$$

where $V_{NO_3}$ and $V_C$ are the measured rates of nitrate and photosynthetic carbon fixation respectively, and *POC* and *PON* are the measured particulate organic carbon and organic nitrogen concentrations, with all values integrated from the surface to the depth of 0.1% light penetration. The resultant estimates for the Ross Sea ice-edge bloom ranged from 35 to 93%, with a mean of 64%. The actual extent of new production is possibly underestimated by this approach because some of the ammonium may enter the inshore Ross Sea as meltwater from ice shelves and glacial ice (see Chapter 2). These values, however, are still very high and imply that the ice-edge blooms may be important not only as a major contribution to total primary productivity in the Southern Ocean, but also in terms of the biomass and production that such blooms can sustain at herbivore and higher trophic levels (Nelson & Smith, 1986). A similarly high level of new (nitrate-dependent) production has been measured in the marginal ice zone off East Antarctica (see Table 5.2).

### 4.4.3  Bacterial heterotrophy

Tritiated thymidine assays during AMERIEZ indicated that bacterial growth rates were maximal in the top 50 m of the water column that had been recently uncovered by the receding ice edge, and closely followed the distribution of phytoplankton (Fig. 4.7). These rates of bacterial DNA synthesis declined ten-fold to the north of the ice edge, and by a factor of 100 southward of this zone beneath the sea ice. The estimated growth rates all fell within the range of $0.001-0.45$ d$^{-1}$ (mean $0.09$ d$^{-1}$) reported for the Southern Ocean. Dissolved free amino-acid (DFAA) concentrations throughout the AMERIEZ cruise were extremely low (10 nmol l$^{-1}$) with highest values at a site of high phytoplankton biomass.

### 4.4.4  Reproductive dynamics of protozoans

Two tintinnid genera, *Cymatocylis* and *Laackmanniella* were examined in detail during the AMERIEZ cruise (Heinbokel & Coats, 1984). Similar results were obtained for both genera. The proportion of dividing cells rarely exceeded 30%, cf. typically 40–60% in temperate estuarine environments.

Cellular division was not correlated with time of day, ambient chlorophyll *a* concentration or primary production. However, division frequency for *Laackmanniella* was positively correlated with latitude – the highest proportions of dividing cells were recorded at the more southern stations, while *Cymatocylis* division frequency showed the inverse pattern with the highest % of dividing cells at stations well to the north of the ice edge.

## 4.5  Trophic structure

The marginal ice zone is a highly transient region of elevated microbial activity, but it is not yet fully understood how efficiently this carbon and energy flux is transferred to higher trophic levels. Some pelagic organisms of the Southern Ocean have life strategies that might allow them to exploit these brief episodes of production (Deacon, 1982; Tranter, 1982). However, much of this ice-edge biomass may sediment to the sea floor where it supports the large benthic community that lives on the antarctic continental shelf, or contributes to the highly siliceous sediments in the deep basins of the region (see Chapter 6).

A broad comparison of data from the AMERIEZ cruise in November/December 1983 illustrates the strong spatial coupling between organisms from many trophic levels across the marginal-ice zone (Fig. 4.7). This band of high animal biomass and production is also characterised by changes in

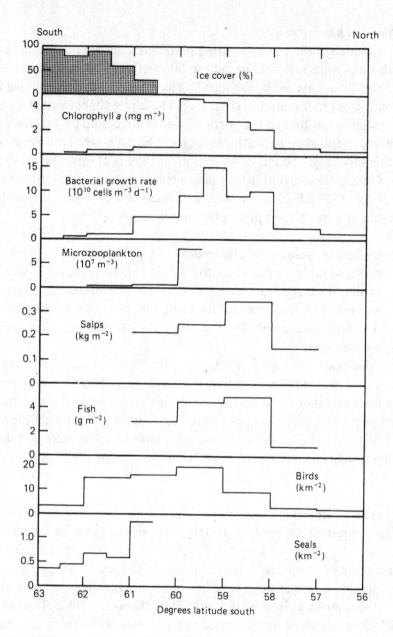

Fig. 4.7. The coupling between algal production and higher trophic levels in the marginal-ice zone. Sampling was conducted in November 1983 along a north–south transect in the Weddell Sea. The data have been collated from: ice cover – Nelson *et al.* (1984) and Comiso & Sullivan (1986); chlorophyll *a* (near surface) – Nelson *et al.* (1984); bacterial growth rates (near surface) – Miller *et al.* (1984); microzooplankton (upper 10 m) – Garrison *et al.* (1984); fishes and salps (upper 1000 m) – Torres *et al.* (1984); birds – Fraser & Ainley (1986); and seals (aerial census) – Erickson (1984).

species composition. For example, there is a transition between two distinct communities of seabirds (Fraser & Ainley, 1986). The pack-ice community is dominated in terms of biomass by penguins (in particular emperor penguins, *Aptenodytes forsteri*, and Adelie penguins, *Pygoscelis adeliae*), but in terms of population abundance by snow petrels (*Pagodroma nivea*) and antarctic petrels (*Thalassoica antarctica*). North of the ice edge, but within the region containing icebergs (less than 20% ice cover) the community shifts to biomass dominance by southern fulmars (*Fulmarus glacialoides*) and cape pigeons (*Daption capense*), with associated populations of Wilson's storm petrels (*Oceanites oceanicus*), mottled petrels (*Pterodroma inexpectata*) and antarctic petrels. The highest population densities of the pack ice community occur about 7–10 km north of the ice edge. This peak in abundance appears to follow the ice retreat at a slight lag. These observations suggest that seabird distribution is not simply dictated by immediate access from the ice edge to open water, but rather is keyed to the surge in marine production that follows (at a lag) the disintegration and melting of the sea ice.

Large differences have been observed in the prey items of seabirds in the two communities which further reflect the differences at lower trophic levels. In the Weddell Sea, north of the ice edge, the main food sources were found to be myctophid fishes (*Electrona antarctica* and *Gymnoscopelus braueri*) and krill (*Euphausia superba*), but within the pack ice the seabird diets shifted to crustacea of the genera *Pasiphaea* and *Eurythenes* (Ainley *et al.*, 1986).

Decaying ice floes may operate as important focal points for all trophic levels within this system. The surface, interior and bottom region of floes support luxuriant growth of algal, bacterial and protozoan communities. Krill and large decapods (*Grennados* sp.) have been observed to enter the anastamosing channel system that permeates the ice floes. Ainley & Sullivan (1984) suggest that this decomposing ice may act as an important island-like refuge for both the floral and faunal components of the ice-edge ecosystem. Certain crustacean species that are *mesopelagic* in the open waters of the ocean (i.e. occur in the depth range 100–1000 m) also occur in the ice channels of these old, decaying floes. These animals are therefore only available to near-surface predators such as seabirds within this pack-ice region (Ainley *et al.* 1986), which further distinguishes the marginal-ice zone from the surrounding ocean.

# 5

## The open ocean

### 5.1　Introduction

The Southern Ocean is a cold deepwater environment that covers about 8% of the total world ocean area. Although many of its basic environmental properties have been known for more than fifty years, knowledge of its food web organisation and production dynamics is continuing to develop rapidly. Early investigators envisaged a highly productive phytoplankton community coupled through a short direct food chain to the prolific higher trophic levels. The open waters of the Southern Ocean are now recognised as a low biomass system with a complex trophic structure. Bacterial heterotrophy, once considered negligible in antarctic waters is now emerging as an important pathway of secondary production, and detrital material, particularly that derived from krill appears to enter the food web at many points. The early expeditions used coarse nets for routine sampling which collected only the large-celled algae (>35 $\mu$m). It is now clear that nanoplankton (2–20 $\mu$m) and picoplankton (<2 $\mu$m) can also be important constituents of the plankton in terms of biomass as well as productivity. The carbon and energy in these small-celled fractions may flow to higher trophic levels (including krill) directly or via grazing by protozoa. These microbial zooplankton include choanoflagellates and tintinnids, and in addition to their trophic role they have also been implicated as important agents of nutrient regeneration.

Diatoms are generally the phototrophic dominants in this nutrient-rich oceanic environment. Three factors appear to keep their productivity and standing crop in check. Low light through deep mixing favours a shade-adapted population of low photosynthetic capacity intermittently exposed to optimal light. The extremely low temperatures (by oceanic standards) appear to substantially depress growth rates. Finally, herbivores in the Southern Ocean such as krill, copepods and salps can achieve extremely high biomass levels and this grazing pressure further limits the population build-up of phytoplankton cells.

## 5.2    The environment
### 5.2.1    *Bathymetry*

The Southern Ocean contains three deep basins (4000–6500 m) delimited to the north by tall ridges and the Kerguelen Plateau (Fig. 5.1). These northern boundaries restrict the free movement of bottom water and also modify the surface currents. Circulation is further affected by the Drake Passage between the Antarctic Peninsula and South America which form a major constriction to circumpolar currents. The South Shetland Archipelago extends the spine of the peninsula northwards to 61 °S and the distance between the 500 m isobaths across the Drake Passage is only 420 nautical miles (about 780 km). The microbial ecology of this constricted region is especially interesting because the distinctive antarctic water masses and fronts are compressed into a narrow zone and large differences in the physical and chemical environment can be observed over relatively small distances.

Antarctica is surrounded by an unusually deep continental shelf. The 'shelf-break', the transition between the continental shelf and continental slope, lies two to four times deeper than in other oceanic regions. In part this may have resulted from the isostatic equilibrium adjustment of the continent to the large mass of ice which covers it. Locally, however, the shelf has been carved and deepened by ice-scouring. Both the Weddell Sea and Ross Sea lie over very broad shelves that have been deepened to 400–500 m by the Filchner Ice Shelf and Ross Ice Shelf, respectively. The inshore microbial populations of the antarctic marine environment may therefore experience deep water more typical of open ocean conditions.

### 5.2.2    *Currents and mixing*

The distribution and activity of microflora in the Southern Ocean are strongly influenced by large-scale hydrographic processes in the region. Unlike the Arctic Basin which is surrounded by landmasses that modify its climate and physical oceanography, the south polar region consists of a huge ice-covered continent around which flow the currents of the open sea.

Two wind-driven currents dominate the surface hydrodynamics of the Southern Ocean (Fig. 5.2). The smaller of these, called *East Wind Drift* or *Antarctic Coastal Current*, flows in an anticlockwise (east to west) direction around the margins of the continent. This current is driven by easterly and southeasterly winds blowing off the continent, and by winds from the northeast and east associated with the southern part of depressions that travel around the continent (see Appendix 1). The average speed of the current is 13–20 cm s$^{-1}$, but rather than forming a continuous band along the coast there is some recirculation in a series of gyres. This current is

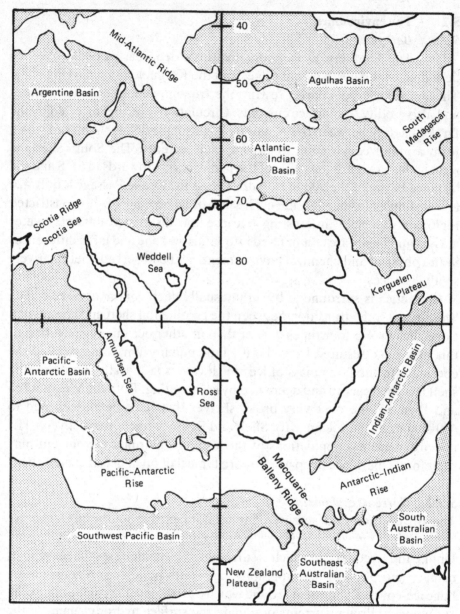

Fig. 5.1. The deep oceanic basins surrounding Antarctica. The shaded areas represent depths greater than 4000 m.

mostly confined to the continental shelf and slope, but in the southeast Weddell Sea it broadens out up to 300 km wide. On the other side of the continent one branch of the current passes outside the entrance to the Ross Sea, while another penetrates southward to flow under the Ross Ice Shelf, returning as a northerly flow along the coast of Victoria Land.

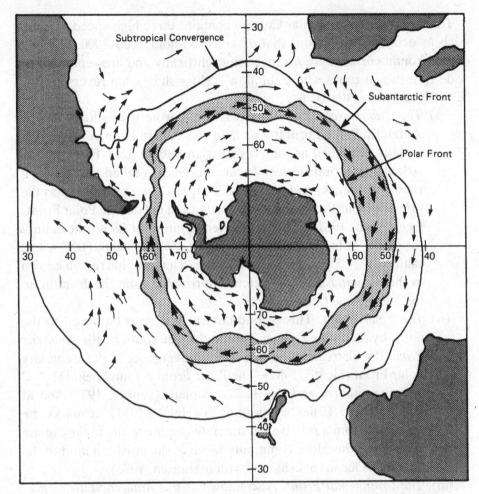

Fig. 5.2. The distribution and flow of surface water masses of the Southern Ocean. The Polar Frontal Zone (shaded) is bordered to the north by the Subantarctic Zone and to the south by the Antarctic Zone. A reverse, anticlockwise current flows in the Continental Zone which lies in the immediate vicinity of the coast. Modified from Heath (1981), Clifford (1983) and Deacon (1984).

Outside the coastal current and north of about 60 °S the strong westerly winds generate a vast clockwise flow called the *West Wind Drift* or *Antarctic Circumpolar Current*. This current is relatively slow (15–20 cm s$^{-1}$, compared with more than 200 cm s$^{-1}$ in the Gulf Stream, for example), but the flow extends to the sea floor and more water is transported than in any other oceanic current system. The average mass flow by this system through the Drake Passage has been estimated at about $1 \times 10^8$–$1.75 \times 10^8$ m$^3$ s$^{-1}$ (Bryden & Pillsbury, 1977).

The Antarctic Circumpolar Current contains three high-speed currents each associated with a front that extends to about 1000–2000 m depth. These frontal currents are arranged concentrically and are separated by wide transitional zones where the flow is sluggish or even reversed (Fig. 5.2). The frontal systems are:

(i) *The Subantarctic Front.* This is the northern boundary of the Antarctic Polar Frontal Zone. In the Australasian sector it was initially described as the Australasian Subantarctic Front; in the Atlantic and southwestern Indian Oceans it is called the Subtropical Convergence, and in the Pacific it has been referred to as the primary Polar Front, the more northern of the double Polar Fronts (see Heath, 1981). The surface temperature at its northern limit (about 52 °S, for example, at longitude 160 °E) is about 8 °C with a salinity of 34.5‰ (Emery, 1977). An oceanographic region known as the *Subantarctic Zone* extends northward from this front up to the *Subtropical Convergence.*

(ii) *The Polar Front.* This second current system is bordered to the north by the *Polar Frontal Zone* and to the south by the *Antarctic Zone*. It is also called the Antarctic Convergence or the secondary Polar Front (Heath, 1981). The Polar Front is found near the 2 °C surface isotherm and the 34.2‰ isopleth (Emery, 1977) and at longitude 160 °E lies at about 56 °S (Heath, 1981). It marks the transition from a relatively temperate climate to the polar climate zone and is therefore commonly taken as the northern limit to the Southern Ocean (Sakshaug & Holm-Hansen, 1983).

(iii) *The Continental Front.* Also known as the *Antarctic Divergence*, this current is separated from the Antarctic coast by the *Continental Zone* which is seasonally covered by sea ice and is the region of the East Wind Drift.

These various currents vary in their strength and location. For example, in the Drake Passage the Polar Front meanders up to 100 km north and south of its mean path. These meanders can pinch off to form cyclonic rings containing cold core water (e.g. Joyce, Patterson & Millard, 1981).

Oceanographers further distinguish the major water masses of the Southern Ocean on the basis of salinity, temperature and the origins of the water (Fig. 5.3):

(i) *Antarctic Intermediate Water.* This forms by mixing across the Subantarctic Front, although the mechanisms of transport are still poorly defined. The water is more dense than water to the north, and less dense than water to the south. It therefore moves northwards and sinks, ultimately to about 1000 m.

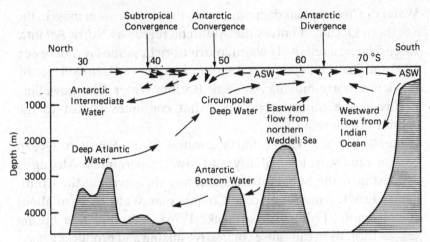

Fig. 5.3. Vertical distribution and flow of water masses in the Southern Ocean along longitude 30° W. Modified from Deacon (1984). ASW = Antarctic Surface Water.

(ii) *Antarctic Bottom Water.* This water mass is primarily formed as a mixture of Antarctic Circumpolar Water and Antarctic Surface Water that has slightly increased in density by winter freezing and then sinks along the continental slope. It is primarily formed in the Weddell Sea, at about $2 \times 10^6$–$5 \times 10^6$ m$^3$ s$^{-1}$. Slightly saltier bottom water is formed in the Ross Sea, but at about one-tenth of this production rate (Foster, 1984). Coastal *polynyas* (persistent ice-free areas within the pack ice) appear to play a major role in the formation of this highly saline water (Zwally, Comiso & Gordon, 1985). Large quantities of sea ice are formed in such areas (e.g. potentially up to 60 m y$^{-1}$ in the Terra Nova Bay polynya) but are continuously removed by the wind. Concentrated brines are produced during the freezing process and these sink to the sea floor to form high-salinity shelf water that is a precursor to Antarctic Bottom Water (Fig. 4.3).

Antarctic Bottom Water is highly saline (Table 5.1) and is the most dense water in the free ocean. It flows northward as a density current tracing the bottom topography of the Southern Ocean, and ultimately feeds into the great oceanic basins. The influence of this water mass can be seen as far north as 40 °N in the Atlantic and 50 °N in the Pacific Oceans. This dense bottom water, generated in the Antarctic, is responsible for the low temperatures (−0.5–2 °C) at great depth throughout world's oceans (Tchernia, 1980).

(iii) *Antarctic Circumpolar Water.* Also known as Antarctic Warm

Water or Circumpolar deep water, it is the largest water mass in the Southern Ocean. It enters the Antarctic region as North Atlantic Deep Water, a relatively warm, nutrient-rich, saline layer between 1000 and 2000 m depth. It is then forced up towards the surface by the dense, north-moving Antarctic Bottom Water. This upwelling begins at the Subantarctic Front but continues almost to the continental margin.

(iv) *Antarctic Surface Waters.* Surface waters of the Southern Ocean have a relatively low salinity and low temperatures. Mixing is deepest near the Subantarctic Front and shallowest at the Continental Front, where Antarctic Circumpolar Water rises to about 100 m depth (Foster & Carmack, 1976). Salt rejection during sea-ice formation can cause convective mixing and produces a deep mixed layer with a salinity as low as 33.5‰. Local melting of sea ice and icebergs in summer lowers the salinity of the near-surface waters and can result in a stable, stratified water column. The lower strata are the cold (near freezing, like the overlying surface waters), more saline (approaching 34.5‰) remnants of winter convection and are referred to as *Winter Water*.

Table 5.1. *Temperature and salinity characteristics of the major water masses in the Southern Ocean, listed by increasing density*

| Water mass | Temperature (°C) | Salinity (‰) | Seawater density $(\sigma_T)^a$ |
|---|---|---|---|
| *Antarctic Intermediate Water* | 2.0 | 33.80 | 27.03 |
| *Antarctic Surface Water* | | | |
| summer | −0.5 to −1.8 in CZ,[b] up to 6 °C in PFZ[c] | 32–34 | – |
| winter | −1.85 to −1.88 in CZ, up to 3 °C in PFZ | 34–34.5 | 27.60 |
| *Antarctic Circumpolar Water* | 0.5–2.5 | 34.68–34.76 | 27.80–27.84 |
| *Antarctic Bottom Water* | −0.4 | 34.66 | 27.88 |

[a] $\sigma_T = 1000 \, (\varrho_T - 1)$, where $\varrho_T$ is the density of water in g cm$^{-3}$ at ambient temperature ($T$) and 1.0 atm pressure.
[b] CZ = Continental zone.
[c] PFZ = Polar frontal zone.
Compiled from Tchernia (1980).

### 5.2.3 Temperature

The extreme cold of the Southern Ocean limits all microbial processes. Temperatures range from about 5 °C in the Polar Frontal Zone to −2.0 °C in ice-filled waters towards the coast (Table 5.1). These are comparable with Arctic Ocean temperatures which are typically around −1.5 °C, although seasonally much higher (up to 6 °C) in the Subarctic.

These cold temperatures are remarkably constant despite the highly seasonal radiation cycle. South of the Polar Front surface temperatures vary by no more than 3–5 °C annually, and the variation is negligible close to the continent where much of the solar energy in summer is consumed in melting the sea ice. In McMurdo Sound, for example, temperatures range seasonally from −1.7 to −1.9 °C.

### 5.2.4 Light

Integral daily insolation at the top of the atmosphere fluctuates enormously between seasons, but varies over a relatively small range between different locations in the Southern Ocean (Appendix 2). Of greater importance may be the transparency of the atmosphere. Anticyclonic conditions prevail near the coast and the skies are often lightly cloudy or clear. In the Circumpolar Current region there is a continuous passage of low pressure systems and a predominance of cloudy weather (see Appendix 1). Insolation may therefore be lower in this region relative to close to the continent (Holm-Hansen *et al.*, 1977).

The oceanic waters around Antarctica are typically blue and highly transparent. The maximum Secchi depth is about 40 m (Slawyk, 1979) comparable with oligotrophic oceans. The 1% light level lies at about 100 m. Tilzer, Von Bodungen & Smetacek (1985) attribute the low concentrations of dissolved coloured compounds and the low abiotic turbidity of these waters to the minimal terrestrial influence within the region.

Phytoplankton cells appear to be the dominant contributor to organic particulate material (mean of 75% of total particulate organic carbon in the study of Tilzer *et al.*, 1985) and are therefore the dominant light absorbing component of the Southern Ocean. This high fractional absorption of light by algae is comparable with extremely clear oligotrophic waters elsewhere, such as ultra-oligotrophic Lake Tahoe, USA. The water is most transparent to wavelengths less than 500 nm wherever algal concentrations are low, but maximum transparency shifts to wavelengths greater than 500 nm in regions of high chlorophyll *a* concentrations (Tilzer *et al.*, 1985).

Phytoplankton in the Southern Ocean must experience a highly variable light regime associated with vertical mixing processes. The upper water column in this region is generally much more unstable than in temperate, tropical or arctic waters (Sakshaug & Holm-Hansen, 1986), and the resultant deep mixing may severely limit the availability of light for photosynthesis.

### 5.2.5   *Nutrients*

The continuous upwelling in the Southern Ocean results in some of the highest macronutrient levels to be found in surface oceanic waters anywhere in the world. Unlike the Arctic Ocean, nitrogen and phosphorus supplies are never exhausted even after the seasonal bloom. For example in the Scotia Sea, nitrate concentrations are typically about 30 mmol N $m^{-3}$, ammonium 0.1–2.0 mmol N $m^{-3}$, and reactive phosphorus ('phosphate') 2 mmol P $m^{-3}$. Silicate levels are extremely high (up to *c.* 100 mmol Si $m^{-3}$); about eight-fold higher than in the Arctic Ocean relative to nitrate and phosphate. Antarctic diatoms appear to be silicified to a much greater extent than in the Arctic, reflecting this elevated abundance of silica. However, their high demand for this nutrient may drive it to much reduced concentrations in parts of the Southern Ocean, particularly towards the north e.g. <5 mmol Si $m^{-3}$ near the convergence.

. There have been few attempts to measure concentrations of dissolved organic C, N and P in the Southern Ocean. However, as in other parts of the world's oceans dissolved organic nitrogen levels are high relative to nitrate and ammonia, e.g. 2–3 times the $NO_3$ concentration in the inshore Ross Sea (Vincent & Vincent, 1982a).

## 5.3    Microbial communities

### 5.3.1   *Microalgal communities*

*Population structure*

Pennate and centric diatoms dominate the microbial biomass of the Southern Ocean (Fig. 5.4). The main genera among the large-celled forms are *Chaetoceros, Corethron, Nitzschia* and *Thalassiosira.* However, it is increasingly clear that small-celled species (<20 $\mu$m) are very important in many locations. Throughout the East Wind Drift, for example, samples collected during mid- to late-summer were often dominated both in terms of biomass and cell concentration by *Fragilariopsis nana*, a pennate diatom about 5 $\mu$m long (Hewes *et al.*, 1985).

Although most of the antarctic marine phytoplankton are circumpolar and are distributed throughout the Southern Ocean, there can be large regional differences in species composition. For example, in a November–December cruise between 54 and 64 °W, *Corethron* dominated the low

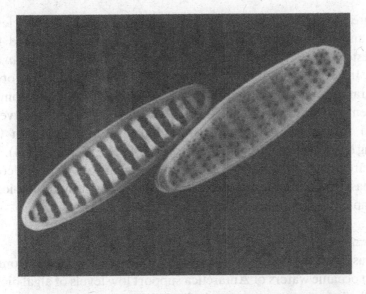

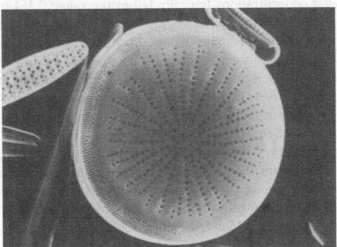

Fig. 5.4. Pennate (e.g. *Nitzschia* (*Fragilariopsis*) *kerguelensis*, above) and centric (e.g. *Actinocyclus actinochilus*, below) diatoms are frequently the microbial dominants in the Southern Ocean. Scanning electron micrographs by Eva Braaten, courtesy of Dr Grethe R. Hasle, Department of Biology, University of Oslo.

biomass assemblage in the southern Scotia Sea; east of Bransfield Strait the community was dominated by *Thalassiosira, Rhizosolenia, Chaetoceros* and *Nitzschia*; and in the high biomass community of the Bransfield Strait the prymnesiophyte *Phaeocystis* as well as the latter four diatoms were the generic dominants (Tilzer *et al.*, 1985). Large floral changes are generally recorded at the Polar Front (El-Sayed, 1984).

Inshore blooms of algae tend to be dominated by a single species. The richest and most extensive of such blooms has been reported in the southwestern Weddell Sea and was composed entirely of *Thalassiosira tumida*. It covered an area of at least 15 000 km$^2$ with a chlorophyll *a* concentration of about 190 mg m$^{-3}$ (El-Sayed, 1971). Blooms composed almost entirely of the colonial species *Phaeocystis pouchetii* have been reported from various locations around the inshore Southern Ocean including the Ross Sea (Vincent & Vincent, 1982a; El-Sayed, 1984). These blooms are not characteristic of the open ocean and are probably controlled, at least in part, by the reduced depth of mixing in the marginal-ice zone (see Chapter 4) and in the shallower inshore waters.

### Biomass structure

In contrast with the highly productive coastal and ice-edge environments the open oceanic waters of Antarctica support low levels of algal biomass. Chlorophyll *a* concentrations typically fall in the range 0.1–1.0 mg m$^{-3}$ (El-Sayed, 1984). Maximum values are usually found well below the surface, with a gradual decrease in concentration towards 200 m and an abrupt decline below this depth. These concentrations are low relative to upwelling systems elsewhere (e.g. >10 mg m$^{-3}$ in the Benguela upwelling off South Africa; Brown & Field, 1986), but are more comparable with nutrient-poor oligotrophic environments (e.g. average of 0.14 mg m$^{-3}$ in the Mediterranean Sea; Azov, 1986).

There is a high level of horizontal variation, with no consistent differences between regions. For example, several studies have identified the Polar Front as a region of particularly low standing crop, and these observations have been attributed to mixing below the euphotic zone (Hart, 1942; El-Sayed & Mandelli, 1965; Hasle, 1969). Conversely, elevated chlorophyll *a* concentrations have been reported in several cruise transects across the Polar Front (e.g. Allanson, Hart & Lutjeharms, 1981) and sedimentation rates for biogenic silica suggest high rates of primary production in this region. As noted above, the surface waters in most parts of the antarctic open ocean tend to be unstable and critical depth effects seem unlikely to account for local changes such as across the Polar Front.

Total microbial biomass in spring ranged from <1–10 $\mu$g C l$^{-1}$ in the Pacific Section of the Southern Ocean and 1–25 $\mu$g C l$^{-1}$ in the Drake Passage (Hanson & Lowery, 1985). These values are comparable with levels in the Northern Pacific Gyre (5–25 $\mu$g C l$^{-1}$). In coastal antarctic waters the values range much higher, up to 1 mg C l$^{-1}$ (Hodson *et al.*, 1981), with correspondingly high levels of chlorophyll *a*. Even the inshore communities become increasingly sparse towards late summer and autumn.

Net plankton and nanoplankton are both important contributors to community biomass and productivity in the Southern Ocean. For example, Hayes, Whitaker & Fogg (1984) reported that north of the Polar Front cells retained by a 35 $\mu$m net made a negligible contribution to the total carbon fixed. South of the front, at 57° 30'S, this large-celled fraction increased in proportional abundance and activity to account for 56% of total C-fixation. The >35 $\mu$m fraction was similarly important (51%) in the Weddell Sea (El-Sayed & Taguchi, 1981).

The relative contribution of each algal size fraction varies greatly with location. In samples from the East Wind Drift, on average 63% of the total chlorophyll *a* passed through a 20 $\mu$m screen. A similar proportion of nanoplankton chlorophyll was recorded in the Weddell Sea (53%), but in the Scotia Ridge area the percentage rose to 86% and many of the samples were dominated by nanoflagellates (Hewes *et al.*, 1985).

### 5.3.2    *Bacterial communities*

Epifluorescence counts of bacteria varied from $1 \times 10^7$–$2 \times 10^8$ cells $l^{-1}$ in the Drake Passage during January, approximately a factor of ten lower than in the Scotia Sea and coastal antarctic waters (Hanson *et al.*, 1983a). Rough estimates of the percentage contribution of bacterial carbon to total microbial carbon ranged from 0.4 to 2.6% in the Drake Passage (Hanson *et al.*, 1983a) to about 10% in McMurdo Sound (Hodson *et al.*, 1981). In a late winter–spring cruise (Sept–Oct) across the Drake Passage bacterial concentrations were 2–3 times higher than in the eastern South Pacific Ocean, with highest values in the Polar Front ($3.5 \times 10^8$ cells $l^{-1}$, Hanson *et al.*, 1983b).

There can also be large local variations in microbial biomass and activity. For example, the western side of McMurdo Sound is highly oligotrophic by comparison with the eastern side, and this is reflected in order of magnitude differences in bacterial concentrations and turnover times for specific organic molecules. On the eastern side of the Sound bacteria averaged $6.5 \times 10^8$ cells $l^{-1}$ and the turnover times for dissolved ATP, D-glucose and L-leucine averaged 16, 116, and 124 h, respectively. On the western side of the Sound the equivalent values were $6.5 \times 10^7$ cells $l^{-1}$, 59 h, 20454 h and 3070 h, respectively, (Hodson *et al.*, 1981). Much higher bacterial growth rates, as estimated by incorporation of tritiated thymidine, have also been recorded in the eastern Sound (Fuhrman & Azam, 1980).

This microbial distribution correlates with other environmental features in the McMurdo Sound region. Surface currents flow southward along the eastern side of the sound bringing with them phytoplankton populations that have developed under open water conditions to the north. On the western side the water flows northwards and is derived, at least partially, from beneath the Ross Ice Shelf where there is inadequate light for phytoplankton photosynthesis and growth.

### 5.3.3    *Protozoan communities*

As in other parts of the world's oceans protozoans may be an important link between small-sized particles and the higher trophic levels.

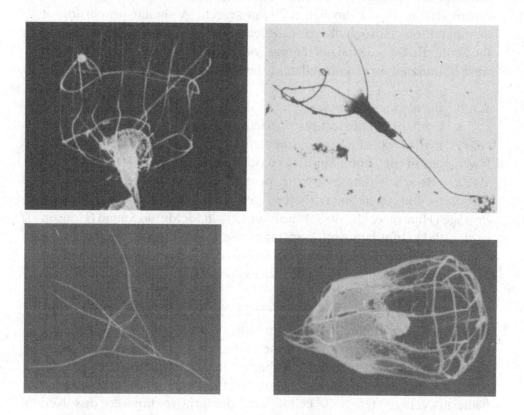

Fig. 5.5. Choanoflagellates appear to be an important link in the Southern Ocean food chain. These species are from the sea off East Antarctica: *Diaphanoeca pedicellata* (top, left), *Bicosta spinifera* (top, right), *Calliacantha simplex* (bottom left) and *Diaphanoeca grandis* (bottom right). The siliceous structures around the cells (lorica) gives these organisms a maximum dimension of about 30–40 $\mu$m. Electron micrographs by Dr Harvey J. Marchant, Antarctic Division, Hobart, Australia.

For 15 stations in the East Wind Drift the biomass of heterotrophic nanoflagellates averaged 24 $\mu$g C l$^{-1}$ (range of 7–60 $\mu$g C l$^{-1}$); this compared with an autotrophic biomass of 48 $\mu$g C l$^{-1}$ (range 10–303 $\mu$g C l$^{-1}$). In the Scotia Sea the total ciliate biomass averaged 3.6 $\mu$g C l$^{-1}$ (range 1.9–7.3 $\mu$g C l$^{-1}$) compared with a mean autotroph (>20 $\mu$m) biomass of 13.9 $\mu$g C l$^{-1}$ (range 0.7–42.6 $\mu$g C l$^{-1}$). Tintinnids accounted on average for about one-third of the total ciliate biomass. Tintinnid and flagellate biomass appeared to be correlated, consistent with the accepted view that nanoflagellates can be food for ciliates (Hewes *et al.*, 1985).

Choanoflagellates appear to be a particularly abundant and conspicuous component of the Southern Ocean plankton (Fig. 5.5). In a detailed study in Prydz Bay this group accounted for 10–40% of the total live nanoplankton cells. The recorded genera in open ocean samples were *Acanthoecopsis*, *Bicosta*, *Calliacantha*, *Crinolina*, *Diaphanoeca*, *Parvicorbicula* and *Pleurasiga*. The overall mean abundance of choanoflagellates averaged 2.7 ($\pm$1.7) $\times$ 10$^4$ cell l$^{-1}$ for the oceanic sites, but ten-fold higher cell concentrations were recorded inshore during the summer (Marchant, 1985).

## 5.4    Microbial processes
### 5.4.1    *Photosynthesis*
Primary production rates in the Southern Ocean typically fall around 0.1 g C m$^{-2}$ d$^{-1}$, comparable with oligotrophic lakes and oceans elsewhere. For example, the geographically wide-ranging Eltanin cruises 38, 46 and 51 gave a mean estimate of 0.134 g C m$^{-2}$ d$^{-1}$ (El-Sayed & Turner, 1977). Much higher values have been recorded in the vicinity of the coast, e.g. 3.2 g C m$^{-2}$ d$^{-1}$ in Gerlache Strait (El-Sayed, 1967) and 2.8 g C m$^{-2}$ near Signy Island (Horne *et al.*, 1969). These inshore photosynthetic rates are comparable with rich upwelling systems (e.g. 2.0 g C m$^{-2}$ d$^{-1}$ off northwest Africa, and up to 3.2 g C m$^{-2}$ d$^{-1}$ off Peru; Barber & Smith, 1981) and probably contributed towards the early and misleadingly high estimates of overall productivity in the Southern Ocean (El-Sayed, 1984).

The photosynthetic profiles typically show a maximum at about the 25–50% surface irradiance level. A considerable portion of the profile may extend below the 1% surface light level which limnologists and oceanographers often consider to represent the bottom of the euphotic zone. In the Weddell and Ross Seas nearly one-quarter of the integral water column production occurred below the 1% level; in the Atlantic and Pacific sectors of the Antarctic these lower depths contributed less than 10% to the total production (El-Sayed, 1984).

The variation in antarctic photosynthesis correlates reasonably well with spatial and temporal variations in phytoplankton standing crop. Both variables track the high amplitude seasonal change in light. The spring bloom is delayed with increasing latitude due to the combined effects of incident irradiance, seasonal ice cover, and associated marginal-ice zone effects. The largely ice-free 'Northern Region' (Polar Front to 600 km [about 330 nautical miles] south) has a peak in early December; the 'Intermediate Region' and Scotia Sea have their peak in January; and the 'Southern Region' (south of the Antarctic Circle, which lies at latitude 66° 33'S) in mid- to late-February (Sakshaug & Holm-Hansen, 1983).

Microalgal growth in the Southern Ocean seems to be extremely slow for many species, and in part may be a feature of algal cells genetically adapted to low temperatures (see Section 5.4.2). Specific growth rate estimates vary from 0.1–0.3 doublings $d^{-1}$ in the Ross Sea (Holm-Hansen *et al.*, 1977), to 0.4–0.6 $d^{-1}$ in the Indian sector (Jacques & Minas, 1981) and 0.71 $d^{-1}$ in the Weddell Sea (El-Sayed & Taguchi, 1981). These estimates, however, are generally biased towards the large, slow-growing diatoms. Small-celled constituents of the phytoplankton may have much faster growth rates, e.g. $>1.0$ $d^{-1}$ for *Fragilariopsis nana* in the absence of grazing pressure (Hewes *et al.*, 1985).

High productivity to the southwest of South Georgia has been attributed to a sharpening of the vertical temperature gradient and the associated stability that retains phytoplankton under favourable light conditions for growth (Hayes *et al.*, 1984). These stability effects have long been argued for various parts of the Southern Ocean (Hart, 1942; Hasle, 1969), most recently in the marginal-ice zone, but they still remain difficult to assess given the unknown rates of vertical mixing throughout the region.

Oceanic production in this region of the world does not appear to be limited by inorganic nutrients. As noted above, nutrient levels are very high and should support a much larger algal population than that observed. Cellular nitrogen : carbon ratios of $<0.1$ (on an atomic basis) indicate nitrogen deficiency, but ratios from the Southern Ocean typically range well above this, up to 0.17. Protein : carbohydrate ratios of $<1.0$ are an additional indicator of nutrient deficiency, but in antarctic samples this ratio has always been above 5.0 (Sakshaug & Holm-Hansen, 1983). Perhaps the most convincing evidence of abundant nutrients relative to planktonic demand in the Southern Ocean has come from shipboard culture experiments with the natural phytoplankton assemblages. These cultures grew exponentially from ambient chlorophyll *a* levels of $<1$ mg m$^{-3}$ to in excess of 40 mg m$^{-3}$ without any enrichment (Sakshaug & Holm-Hansen, 1983).

Enrichment assays with zinc, molybdenum, cobalt, manganese and iron have provided no evidence of trace element deficiency (Jacques, 1983). Similarly, in a series of growth and photosynthetic assays south of the Polar Front from 20–70 °W there was no response to additions of nitrate, phosphate, silicate, trace metals or vitamins (Hayes *et al.*, 1984).

### 5.4.2 *Phototrophic production: the influence of light and temperature*

Antarctic phytoplankton appear to grow at extremely slow rates. Their photosynthetic systems are saturated at low irradiance values, and their maximum photosynthetic rates per unit chlorophyll (assimilation numbers) are amongst the lowest recorded in the world's oceans. These characteristics were originally interpreted as a necessary adaptation to the purported 'shade' or low light environment of the deeply mixed Southern Ocean. Increasingly, however, it has become apparent that these distinc-tive physiological properties are additionally under the control of the low ambient temperatures.

Two antarctic algal cultures (*Nitzschia turgiduloides* and *Chaetoceros* sp.) grown at 5 °C demonstrated a decreasing $I_k$ with decreasing irradiance during growth; but even the highest irradiance cultures had a low $I_k$ (19–26 W m$^{-2}$) and a low assimilation number (0.8–1.4 mg C (mg chl$a$)$^{-1}$ h$^{-1}$). Natural phytoplankton samples from the Indian sector of the Southern Ocean ($c.$ 66 °E) showed a similar effect. $I_k$ values varied between 3 and 10 W m$^{-2}$ (mean of 7.1 W m$^{-2}$) and assimilation numbers were consistently low (1–2 mg C (mg chl$a$)$^{-1}$ h$^{-1}$). There was no definable trend with latitude (Jacques, 1983).

At least some of the planktonic algal species in the Southern Ocean appear to be obligate psychrophiles, i.e. with a growth optimum at low temperatures and an impaired ability to grow under warmer conditions. For example, three diatom isolates from Antarctica were unable to grow above 10 °C, and two of the species had growth optima at 5 °C or below (Fig. 5.6). The maximum growth rates were extremely slow, from 0.45–0.65 doublings d$^{-1}$ (Jacques, 1983). These rates are well below the theoretical maximum predicted by Eppley (1972) and suggest genetic adaptation to low seawater temperatures. Antarctic phytoplankton, however, do not seem to possess adaptive mechanisms that would allow them to grow faster at low temperature. For example the maximum growth rates measured by Sakshaug & Holm-Hansen (1986) were $<0.5$ doublings d$^{-1}$,

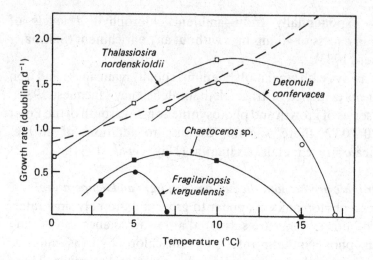

Fig. 5.6. Temperature optima for marine diatoms from the Arctic Ocean (open symbols) and the Southern Ocean (closed symbols). The broken line is the curve derived by Eppley (1972) as the upper bound to growth rates for a wide range of algal culture data. Modified from Jacques (1983).

no higher than the growth rates of temperate species at Southern Ocean temperatures.

Short-term assays of photosynthetic rates suggest somewhat higher temperature optima, but generally confirm the strong temperature dependence of phototrophic production in antarctic waters. The most comprehensive attempt to define the interactive effects of light and temperature on photosynthesis in the Southern Ocean has been a set of controlled incubator experiments performed by Tilzer *et al.* (1986) in the South Scotia Sea. Photosynthesis increased with increasing temperature both at saturating and limiting irradiance levels; the greatest temperature effect was at light saturation. Under light-limiting conditions the $Q_{10}$ (change in rate for a 10 °C rise in temperature, see section 10.2.2) for photosynthetic $CO_2$ fixation was 2.6 over the range $-1.5$–5 °C, but there was no further enhancement above 5 °C. The $Q_{10}$ for light-saturated photosynthesis ranged from 4.6 between $-1.5$ and 0 °C, to 1.3 between 2 and 8 °C. These values for the low temperatures lie above the $Q_{10}$ estimates at lower latitudes (typically in the range 1.8–2.3) but the high level of temperature-responsiveness is consistent with earlier data from the Southern Ocean obtained by Neori & Holm-Hansen (1982) under less controlled conditions.

### 5.4.3 *Phototroph respiratory losses*

An important feature contributing towards the net production of algal biomass in the Southern Ocean may be reduced respiratory losses at low ambient seawater temperatures. Tilzer & Dubinsky (1987) have reported much higher $Q_{10}$ values (2.3–11.9) for phytoplankton respiration than for photosynthesis (1.4–2.2) in the southern Drake Passage and Bransfield Strait. Their measurements need to be repeated using other methodologies to check these rates (respiration is a difficult parameter to quantify in low biomass systems) but they imply that the mass balance of phytoplankton (photosynthetic gains minus respiratory losses) is more positive in the cold ambient temperatures than it would be in warmer environments. Model calculations indicate that these greater effects of temperature on respiration than photosynthesis will be especially critical to net production during the short day lengths or deep mixing regimes that operate throughout much of the year (Fig. 5.7).

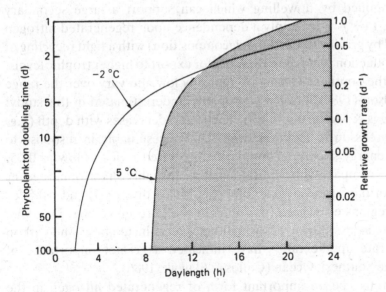

Fig. 5.7. The dependence of algal growth on daylength assuming the only loss is night-time respiration. These curves have been derived from the relationship: $k$ (net specific growth rate) $= \mu^* t_L - p t_D$, where $t_L$ and $t_D$ are the number of hours of light and dark respectively, and $p$ is the respiratory loss rate and $\mu^*$ the photosynthetic rate measured at a Southern Ocean station. Redrawn from Tilzer & Dubinsky (1987).

### 5.4.4  *Phototrophic Nutrition*

Inorganic nitrogen supply is often considered a limiting factor for microalgal growth in the world's oceans. In the seas surrounding Antarctica nitrate is present in abundant supply, and complete depletion of this nutrient seems highly unlikely. Despite these $NO_3$-replete conditions, more than 50% of the nitrogen assimilated by antarctic phytoplankton often appears to be in the form of ammonia. This 'regenerative' type of community nutrition in a 'vast sea of nitrate' is possibly a response to the low energy conditions that the phytoplankton experience during deep mixing (Sakshaug & Holm-Hansen, 1983), but methodological problems have also been raised that suggest the importance of nitrate could be greater than the current data indicate (Collos & Slawyk, 1986).

The relative uptake of nitrate and ammonium ($f$) is commonly expressed in biological oceanographic studies by the ratio:

$$f = V_{NO_3}/(V_{NH_4} + V_{NO_3})$$

where $V$ is the uptake rate of nitrate or ammonium measured with $^{15}N$-labelled tracers. High $f$-ratios indicate a reliance upon so-called 'new' nitrogen supplied by upwelling which can support a large secondary production. Low $f$-ratios imply a dependence upon regenerated nitrogen (N released by grazing or bacterial decomposition) with a tight recycling of primary production that is not available for export to higher trophic levels. $F$-ratios in the Southern Ocean are typically low and vary over the range 0.2–0.5 (Olson (1980), but $f$ was probably underestimated in this study; Gilbert, Biggs & McCarthy, 1982). The $f$-ratio decreases with depth (e.g. Ronner, Sorensson & Holm-Hansen, 1983) presumably in response to decreasing energy supply, giving depth-integrated ratios as low as 0.05. These values compare with $f$-ratios of 0.1–0.2 in highly oligotrophic oceans (e.g. 0.16 for the Sargasso Sea, Gilbert & McCarthy, 1984) and >0.5 for upwelling regions elsewhere (e.g. 0.66 in the Peru upwelling, Dugdale, 1976). Other assays, however, using longer term incubations and carbon uptake: nitrate uptake ratios have indicated a greater importance of nitrate in the Southern Ocean (Collos & Slawyk, 1986).

Urea can also be an important form of regenerated nitrogen in the Southern Ocean (Probyn & Painting, 1985). In a series of experiments off the antarctic coast between Cape Ann and Mawson urea was detected in concentrations that were comparable with or greater than ammonium concentrations (Table 5.2). $F$-ratios, calculated as $V_{NO_3}/V_{\Sigma N}$ where $V_{\Sigma N}$ is the sum of the uptake rates for ammonium, nitrate and urea, ranged from 0.2–0.5. The contribution of urea to $V_{\Sigma N}$ varied greatly with site, but was sometimes comparable with or higher than $V_{NH_4}$. A relative preference index was calculated for each of the three nitrogenous nutrients, for example for ammonium as:

$$(V_{NH_4}/V_{\Sigma N})/([NH_4]/[\Sigma N])$$

where $[NH_4]$ and $[\Sigma N]$ are the measured concentrations of ammonium, and of the three forms of nitrogen, respectively. This index thus compares the relative rate of uptake of a form of nitrogen with its relative abundance. The index for both urea and ammonium were always well above 1.0 indicating a marked preference by antarctic phytoplankton for these two substrates. The index for urea was often similar to that for ammonium while the index for nitrate was always very low, 0.6 or less. In parts of the Southern Ocean then, urea is taken up in preference to nitrate and may be a substantial contributor to phytoplankton nutrition.

The various size components of the phytoplankton community in the Southern Ocean differ in their contribution towards total nitrogen uptake. In the study off East Antarctica by Probyn & Painting (1985) the plankton populations were dominated (in terms of nitrogen biomass) by cells less than 15 $\mu$m. Picoplankton (<1 $\mu$m) accounted for more than half of the total plankton biomass both at sites near the ice edge and 110 km further north in the open ocean (Table 5.2). At open ocean sites picoplankton accounted for the largest fraction of nitrogen uptake but at the ice edge net plankton (15–200 $\mu$m) became more important and were responsible for about half of the total nitrogen uptake. Nanoplankton and picoplankton at both sites discriminated against nitrate and showed a marked preference for ammonium and urea. At the ice-edge site a large percentage (71%) of the nitrate taken up went into the net plankton, but in the open ocean this large size-fraction made a negligible contribution to nitrate uptake. These ice-edge data translate into a very high $f$-ratio for the net plankton (0.9) by comparison with the nanoplankton (0.4) and picoplankton (0.3) suggesting that larger cells may dominate the new production that fuels higher trophic levels in the marginal-ice zone (see Chapter 4).

Antarctic diatoms appear to have an unusually high demand for silicate. In the Indian sector the uptake ratios were 6.0 for Si : N and 88 for Si : P (LeJehan, 1982). This effect was seen in the particulate Si : N ratios (atomic) which increased from 0.5 in the subtropics to 2.0 near the polar

Table 5.2. *The importance of picoplankton (cells <1 μm) in nitrogen uptake in the Southern Ocean*

The two plankton communities were sampled in March–April near latitude 65 °S and longitude 60 °E. The 'total' community was prefiltered through a coarse net and represents cells <200 μm. The contribution of the picoplankton to the particulates or N-uptake is expressed as a percentage of the 'total'.

| | Ice-edge station[a] | | Open ocean station[b] | |
|---|---|---|---|---|
| | 'Total' | Picoplankton | 'Total' | Picoplankton |
| *Particulates* (mmol m$^{-3}$) | | | | |
| organic-C | 7.56 | 73% | 4.92 | 68% |
| organic-N | 0.71 | 60% | 0.53 | 63% |
| *N-uptake* ($10^{-3}$ mmol m$^{-3}$ h$^{-1}$) | | | | |
| NH$_4$ | 1.57 | 54% | 0.75 | 92% |
| Urea | 0.69 | 65% | 0.22 | 59% |
| NO$_3$ | 3.65 | 15% | 1.15 | 84% |
| *Relative preference index* | | | | |
| NH$_4$ | 20.8 | 36.7 | 8.5 | 39.2 |
| Urea | 5.0 | 10.7 | 11.8 | 7.8 |
| NO$_3$ | 0.6 | 0.3 | 0.6 | 0.6 |

[a]Station 30; NH$_4$ = 0.3, Urea = 0.6, NO$_3$ = 25.6 mmol N m$^{-3}$.
[b]Station 46; NH$_4$ = 0.2, Urea = 0.2, NO$_3$ = 23.9 mmol N m$^{-3}$.
Compiled from Probyn & Painting, 1985.

front and 2.45 at 56 °S (Copin-Montegut & Copin-Montegut, 1978). Similarly, isolates of large antarctic diatoms in culture require very high concentrations of silicate to achieve their maximum uptake rate for this substrate. Their half saturation constants (K$_s$ values) are a factor of 2–4 above most other species; e.g. a K$_s$ of 12 mmol Si m$^{-3}$ for *Fragilariopsis kerguelensis* (Jacques, 1983).

### 5.4.5   *Bacterial heterotrophy*

Our understanding of bacterial processes in the Southern Ocean has changed radically over the last few decades. Early investigators concluded that bacteria would be strongly inhibited by low temperatures and that dissolved organic matter (DOM) oxidation rates would therefore be extremely slow. Sorokin (1971) hypothesised that this high residual DOM would be advected to low latitudes by deep oceanic circulation, and eventually upwelled to contribute towards productivity in the tropics. With the application of new bacterial enumeration and assay techniques it has become clear that antarctic surface waters contain a wide range of active bacteria, that their abundance is comparable with temperate latitudes and that the bacterial turnover of DOM can be fairly rapid.

Like certain phytoplankton of the region many of the Southern Ocean bacteria are believed to be obligate psychrophiles with optimal growth below 10 °C. Organic substrate uptake in McMurdo Sound increased by about 30% between −1.8 °C (ambient seawater temperature) and 5 °C, but was increasingly depressed by temperatures in the range 5–25 °C (Hodson *et al.*, 1981). Wiebe & Hendricks (1974) report that 80% of their bacterial isolates from the Southern Ocean did not grow above 10 °C. Temperature rather than organic substrate availability appears to limit bacterial growth; bacterial samples from the Drake Passage failed to respond to organic enrichment at 0 °C or 5 °C, but showed a positive growth response at temperatures well above ambient.

Despite low seawater temperatures the microbial populations maintain a high energy status indicative of active metabolism. Populations in the Drake Passage had an adenylate energy charge of up to 0.8, slightly higher than populations in the South Pacific section (0.6–0.7). Rates of RNA synthesis (estimated by tritiated adenine incorporation) in the Drake Passage were maximal in the surface waters (0–100 m) with highest rates often in the oxygen maximum/temperature minimum layer (Fig. 5.8). Rates of synthesis dropped up to 1000-fold with increasing depth down to 3000 m (Hanson *et al.*, 1983a). An analysis of particulate ATP levels in the Drake Passage showed that most of the plankton biomass was in the fraction $>3 \mu m$, but most of the RNA synthesis was due to cells $<3 \mu m$. The specific rates of RNA synthesis per unit ATP were extremely low: $0.2 \times 10^{-3}$–$25 \times 10^{-3}$ pmoles (ng ATP)$^{-1}$ d$^{-1}$ which is about 1000-times lower than the specific rates measured in the Pacific Ocean.

Rates of bacterial DNA synthesis (measured by tritiated thymidine uptake and incorporation) were lower in the Drake Passage (0.005–5.4 pmol l$^{-1}$ d$^{-1}$) than in temperate coastal waters (2.4–502 pmol l$^{-1}$ d$^{-1}$, Fuhrman & Azam, 1980). However, the specific rates of synthesis (about $10^{-21}$ moles of thymidine cell$^{-1}$ h$^{-1}$) were comparable with many other marine environments.

Much higher estimates of bacterial production in the Southern Ocean

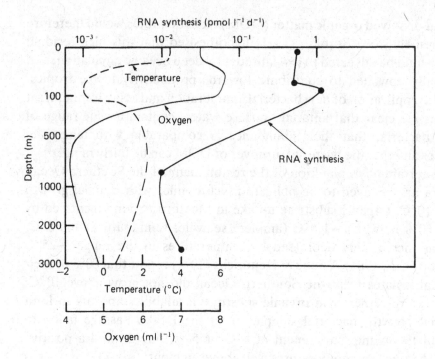

Fig. 5.8. Depth profile of bacterial activity as measured by adenine incorporation into RNA, in the Drake Passage. Maximum activity is located within the oxygen maximum. Modified from Hanson *et al.* (1983a).

have been obtained using an alternative method based on the frequency of dividing cells (FDC). Growth rates in the Drake Passage and South Pacific sector ranged from 0.088–0.032 $h^{-1}$ giving generation times of 11–31 h. The fastest growth rates in the Eastern Pacific sector were at the Polar Front and in the northern Subantarctic Zone, but in the Drake Passage the highest rates were in the vicinity of the sea ice (Hanson *et al.*, 1983b).

From estimates of FDC-derived growth rates and of the standing stock of bacterial carbon (epifluorescence counts) bacterial production rates were calculated to be 2.6–17.1 $\mu$g C $l^{-1}$ $d^{-1}$ in the Drake Passage. These fall at the low end of the range so far reported for marine waters using FDC, but translate into 15–45% of the primary productivity (Hanson *et al.*, 1983b). However, a number of major uncertainties plague the FDC method and its calibration and conversion to *in situ* growth rates, and these estimates of bacterial productivity need to be verified.

As in most oceanic waters, microbial uptake of dissolved organic carbon in the Southern Ocean is mainly due to free-living bacteria. On average 84% of the turnover of dissolved ATP in McMurdo Sound was attributable to cells passing through a 0.6 $\mu$m pore size filter (Hodson *et al.*, 1981). This fraction also dominated the tritiated thymidine uptake, and mostly consisted of free-living cells (Fuhrman & Azam, 1980). In the surface waters of the Drake Passage more than 90% of the measured tritiated glucose flux was associated with the fraction passing through a 3 $\mu$m filter (Hanson *et al.*, 1983a). However, the >3 $\mu$m fraction became increasingly important with depth and by 200 m and below these larger particles dominated the glucose uptake. Similarly the percentage incorporation of tritiated adenine into RNA in the >3 $\mu$m fraction increased from 2.3% at 5 m to 65.5% at 850 m. These data suggested that as phytoplankton cells sink from the euphotic zone into deep water they are subsequently colonised by bacteria and become major sites of microbial activity. The glucose flux rates for these deep-water microbial aggregates, however, were less than 10% of rates in the euphotic zone, and rates of incorporation of adenine into RNA dropped 1–3 orders of magnitude with increasing depth down to 3000 m.

In marine and freshwater environments a high percentage, (generally >50–75%) of the planktonic bacteria appear to be metabolically inactive or dormant. In antarctic waters, as elsewhere, there is little or no correlation between bacterial DNA synthesis and bacterial cell concentrations.

### 5.4.6 *Protozoan grazing*

The abundant protozoa in the Southern Ocean gain their carbon and energy supplies from a wide range of sources, including other protozoa. Detrital particles, bacteria and nanoplanktonic algae, for example, are found within the anterior veil (presumed to be involved in the feeding process) of the choanoflagellate *Diaphanoeca multiannulata*, and this species may also be capable of ingesting the mucilage produced by *Phaeocystis pouchetii* (Marchant, 1985).

The role of microzooplankton has been tested experimentally at two sites in the Southern Ocean (Weddell Sea and Scotia Ridge) by prefiltering seawater samples through a 10 $\mu$m screen (to remove most of the protozoa) or a 202 $\mu$m screen (to remove large zooplankton only). The ungrazed (10 $\mu$m-screened) phytoplankton populations had a faster rate of growth, and in Weddell Sea samples a small-celled diatom achieved growth rates in excess of one doubling per day in the absence of the microherbivores (Fig. 5.9).

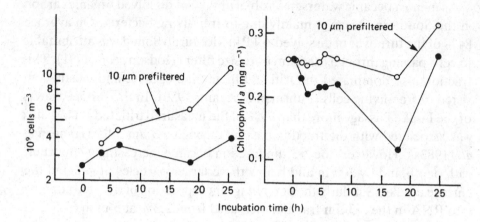

Fig. 5.9. Evidence for microzooplankton grazing. In this experiment water from the Continental Zone was incubated after prefiltration through a 10 or 202 $\mu$m mesh screen. The reduction of chlorophyll *a* in the 202 $\mu$m but not the 10 $\mu$m treatment (right) and the rapid growth of *Fragilariopsis nana* in the 10 $\mu$m prefiltered treatment only (left) suggested strong grazing pressure by heterotrophs in the size range 10–202 $\mu$m. The bottom curves in each graph are for the <10 $\mu$m fraction in the 202 $\mu$m treatment. Redrawn from Hewes *et al.* (1985).

## 5.5    Trophic structure

The classic view of trophic relationships in the Southern Ocean is of a simple three-step food chain that passes from large diatoms to shrimp-like crustacea called krill to vertebrates such as penguins, seals and whales. The krill comprise several euphausiid species, but the most common representative is *Euphausia superba*. These organisms dominate the total animal biomass in the pelagic zone of the Southern Ocean, and fisheries experts predict a sustainable krill harvest of $10^8$ tonnes $y^{-1}$, some 50% above the present total world harvest of fish and shellfish. These estimates of sustainable yield are highly speculative, however, given the uncertainties in krill distribution (e.g. Heywood, Everson & Priddle, 1985) and the sensitivity of this species to food levels at critical stages in its life cycle (Ross & Quetin, 1986).

This enormous biomass of herbivorous krill exerts a strong grazing pressure on the phytoplankton and may be an important factor limiting their standing crop. In the antarctic coastal current, for example, these animals can exceed all other biomass components of the pelagic community by a factor of 10 (e.g. Hopkins, 1985), and phytoplankton often occur in inverse abundance to krill, both over space and time.

Lagged responses of krill to increases in phytoplankton may also explain decreases in algal biomass away from zones of upwelling. The increase in phytoplankton biomass around 65 °S seemed to be associated with the Continental Front, and a delayed response by herbivores as reported in other upwelling systems (e.g. off the Peruvian coast) might favour this local abundance of microalgae (Hayes *et al.*, 1984).

Although there can be no doubt that krill pervade the food web of the Southern Ocean, the trophic structure of antarctic marine communities appears to be more complex than earlier views would suggest. Exchanges take place between the open ocean, the marginal-ice zone and the sea ice, but the mechanisms and magnitude of these interactions remain uncertain. Microbial heterotrophy is now recognised to be an important pathway of carbon and energy transfer. The various protozoan groups such as tintinnids and choanoflagellates may be especially important in transforming particulate material derived from algal and bacterial production into larger 'packages' that are more available to the net zooplankton, including krill. This grazing pressure from protozoa may also be a major constraint on the population size of phototrophs in the pico- and nanoplankton size ranges (Hewes *et al.*, 1985).

Supporting evidence for the trophic pathway through protozoa to krill has come from experiments with *Euphausia superba* in shipboard cultures. The cultures were maintained in running seawater ($-1.0$ to $-0.7$ °C) pumped from the bottom of the ship, and were not supplemented with other food. Analyses of the fecal pellets produced by these euphausiid cultures showed that the animals predominantly fed on the choanoflagellate *Parvicorbicula socialis*. The fecal pellets contained high levels of amino acids, amino sugars and unsaturated fatty acids, and seemed a rich source of nutrients that could be used by heterotrophs as the pellets sank into deeper waters (Tanoue & Hara, 1986). Choanoflagellate grazing may thus play an important role in the transfer of carbon and energy to higher trophic levels, and from the surface to deep-water communities.

Krill appear to enter the food web at many points, as live prey for carnivorous animals, and additionally as debris derived from moults, as detritus, or as body parts incorporated within fecal pellets. In the complex web described by Hopkins (1985) for the surface 100 m of the Croker Passage (Gerlache Strait) krill debris was recorded as a food component for 56 of the 92 observed species of planktonic and small swimming animals. It ranked as the third most important food source, after non-diatom algae and detritus (78 species) and phytoplankton of the *Coscinodiscus* group (64 species). By contrast, *Euphausia superba* was recorded as live prey for only eight species. It is still not known how krill, living and dead, can enter the diets of so many species, but we can be sure that the classic trophic pathway described earlier this century represents only part of total energy flow in the Southern Ocean.

# 6

# Benthic marine environments

## 6.1  Introduction

From the first oceanographic expeditions onward marine zoologists have been fascinated by antarctic benthos. One of the earliest benthic investigators, Dr James G. Eights, visited the South Shetlands in 1830 and described a wide range of new animals from the sea floor including a giant isopod (*Glyptonotus*) and a ten-legged sea spider (*Decolopoda*). The latter species was so unlikely that it took some 70 years before Eights's observations were verified and accepted (Dell, 1972). More recent studies have confirmed the unusual features of the antarctic benthos. The prolific infauna of the inshore marine sediments of Antarctica can achieve population densities almost twice as high as in temperate zone embayments, and more than two-thirds of these species are endemic to the region. Other features of the benthic animal communities in Antarctica include gigantism, prolonged longevity, slow growth rates and delayed maturation. Despite their remarkably high biomass levels the productivity of these bottom-dwelling animal communities appears to be low (White, 1984).

Although many of the faunal properties of the antarctic benthos are now well established the microbial elements of this ecosystem remain little studied and poorly understood. With the exception of the shallow water littoral zone, where a range of microalgae co-exist with the larger algal macrophytes, the shelf ecosystem is dominated by heterotrophs and is fuelled by sea-ice algae or by phytoplankton sinking out from the overlying water column. The microbial dominants may be protozoa, particularly foraminifera. These organisms are phagotrophic (particle-consumers), but may also compete directly with bacteria and fungi for dissolved organic substrates.

## 6.2    The environment

### 6.2.1    *Environmental stability*

The antarctic benthic environment encompasses two extremes of stability. The inshore zone experiences large and irregular disturbances by ice that produce an unpredictable habitat for marine life. Below this narrow zone, however, the benthic environment shifts toward a regime of extreme constancy, comparable with such highly stable habitats as caves and the abyssal ocean.

Ice has a disruptive effect on the shallow water environment through two mechanisms. Drifting pack ice scours the sea floor and may completely strip the inshore zone of loose sediments and sessile flora and fauna. In McMurdo Sound, for example, the depth region 0–15 m is essentially devoid of sessile organisms (Dayton *et al.*, 1974). This physical disturbance is assisted by *anchor ice*, ice crystals which form and grow at the sea floor and which can detach material from the sediments by flotation. The zone of anchor ice extends to 33 m in McMurdo Sound and objects as heavy as 25 kg can be lifted by this mechanism.

Below the anchor-ice zone physical disturbances are rare and the environment is extremely stable. Chemical and physical measurements in deeper water provide further evidence of the low level of environmental variability. Littlepage (1965) reports the following annual means ±2 standard deviations for a large number of observations in McMurdo Sound at 75 m depth: temperature $-1.87 \pm 0.22\,°C$; salinity $34.7 \pm 0.4‰$; dissolved oxygen $9.7 \pm 0.6$ g m$^{-3}$.

### 6.2.2    *Physical substrates*

A wide range of substrata are available for microbial colonisation in the benthic environment and even in one location there may be an enormous variability in community structure. Macroalgae and animals living on the sea floor provide a favourable substrate for some species of microalgae, bacteria and protozoa, but support larger populations of microheterotrophs when they die and begin to decompose. There are large stands of benthic seaweed in many inshore locations. Around the sub-antarctic islands prolific communities of kelp (order Laminariales) offer potentially fertile habitats for microbial activity. South of the Polar Front this group is replaced by members of the related order Desmarestiales. This kelp-like group of brown seaweeds includes the endemic and common antarctic species *Himantothallus grandifolius* which grows as thick undivided blades up to 10 m long and 1 m wide (Moe & Silva, 1977). Dense stands are produced by this species down to about 40 m, but other

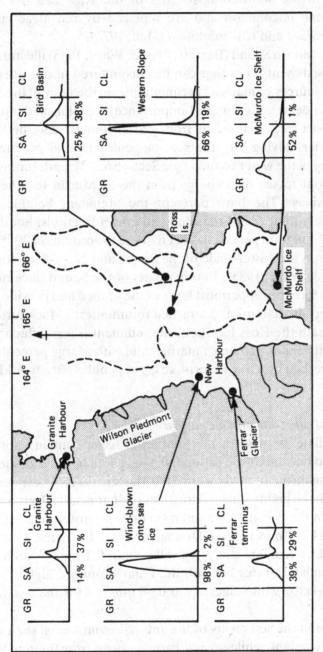

Fig. 6.1. Distribution of particle sizes in various sediment types in the McMurdo Sound region. GR, gravel; SA, sand; SI, silt; CL, clay. Modified from Barrett et al. (1983).

macroalgae such as *Desmarestia menziesii* have been recorded to almost 700 m depth. In the southernmost parts of the Ross Sea the brown seaweeds become uncommon and are replaced by red algae such as *Phyllophora, Iridaea* and *Hildenbrandia* (Dell, 1972).

A study in McMurdo Sound (Barrett, Pyne & Ward, 1983) illustrates the wide range of sediment types that can be encountered in one area (Fig. 6.1). The main sources of material accumulating on the floor of the Sound appear to be (in decreasing order of importance): coastal sand blown by wind onto the sea ice; basal debris from glacier tongues and the Wilson Piedmont Glacier calving into the sea; biogenic material produced by diatoms growing in the water column (see Section 6.2.3); and, supraglacial debris carried out to sea on icebergs from the McMurdo Ice Shelf and ice-cored moraines. The hard parts of the abundant benthic fauna accounted for less than 2% of the sand and gravel fractions, but diatom debris, much of it as fecal pellets, has been found to account for 30–50% of the silt fraction on the eastern side of the Sound and 10–30% on the west (Barrett, Pyne & Ward, 1983). In other parts of the Sound thick mats of sponge spicules form the uppermost layer of the seabed and provide a firm substrate for certain diatoms and attached foraminifera. Much further to the south, beneath the Ross Ice Shelf, the sediment is less influenced by biological activity and is a compact marine mud with a large percentage of glacial flour, overlaid by a thin desert-like lag of pebbles (Brady & Martin, 1979).

### 6.2.3   *Sediment silica and organic carbon*

The sediments beneath the Southern Ocean contain enormous amounts of diatomaceous ooze, estimated to be about 60–80% of the total recent deposition throughout the world (De Master, 1981). These deposits are derived from the highly silicified frustules of diatoms which grow in the surface waters of the ocean. Sedimentation rates in protected shelf areas may be as high as 2–3 mm $y^{-1}$ (see Dunbar, Dehn & Leventer, 1984), but beneath the open ocean are more typically about 0.1 mm $y^{-1}$. Siliceous sediment accumulation rates beneath the Polar Front are highest in the South Atlantic sector with values up to 0.53 mm $y^{-1}$ for the last 18 000 years (Fig. 6.2).

Biogenic silica in the sediments of the antarctic continental shelf range from <1–48% by weight, while organic carbon ranges from 0.1% to nearly 2%. In general the sediments with high levels of biogenic silica are also enriched in organic carbon (Dunbar *et al.*, 1984). The main regions of accumulation of siliceous, organic-rich sediments are between 500 and

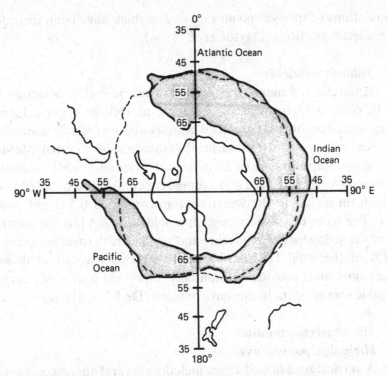

Fig. 6.2. The distribution of the major deposits of siliceous sediments beneath the Southern Ocean. The dashed line indicates the approximate position of the Subantarctic Front. Modified from De Master (1981).

1000 m depth in the many deep basins and glacial troughs of the antarctic shelf. Moderately siliceous sediments are also accumulating at depths greater than 1000 m in the Bransfield Strait basins of the Antarctic Peninsula but biogenic inputs there are obscured by a substantial supply of terrigenous material. At depths on the shelf less than 300 m the sediments are broken down and removed by bottom currents and the biogenic silica content rarely accumulates above 5–10% (Dunbar *et al.*, 1984).

A different type of siliceous environment occurs in certain near-shore regions and provides the habitat for a distinctive microflora. Siliceous sponge spicule mats up to several centimetres thick cover large areas of the sediments at many antarctic sites including shallow areas around King George Island, the Lassiter Coast of the Weddell Sea and Ross Island in McMurdo Sound (Palmisano *et al.*, 1985c). At Cape Armitage, Ross Island, the mats range from 100 mm thickness at 20 m depth to 10 mm at 30 m depth, with abundant siliceous spicules several centimetres in length.

Accumulations of sponge spicules up to 2 m thick have been recorded at some antarctic locations (Dayton *et al.*, 1974).

### 6.2.4    Sediment nutrients

Antarctic sediments lie beneath a nutrient-rich ocean and are likely to contain high levels of inorganic as well as organic forms of nitrogen and phosphorus. Dissolved primary amines (mostly amino acids) have been measured in the surficial sediments of western McMurdo Sound, at concentrations from 22–60 $\mu$mol l$^{-1}$ (mean 35 $\mu$M). Seawater at the sediment–water interface had average concentrations of 6.8 $\mu$M, while samples from higher in the water column contained 0.5–1 $\mu$M primary amines. The main dissolved free amino acids (DFAA) in the interstitial water of the sediments were glycine and serine which together accounted for 57% of the total DFAA. These high concentrations of dissolved organic compounds provide a potentially important source of energy for the benthic foraminifera in this environment (De Laca, 1982a).

## 6.3    Microbial communities

### 6.3.1    Microalgal populations

A vertical zonation of algae, including several microscopic species, is observed on many antarctic shores. The filamentous chlorophytes *Urospora penicilliformis* and *Ulothrix australis* form a green band on rocky shorelines at the high tide level on the Antarctic Peninsula and on offshore islands (Heywood & Whittaker, 1984). The calcareous red algae *Lithophyllum aequabile* and *Lithothamnion granuliferum* form crusts on exposed rocks. Rock pools are often coated with a thick felt of diatoms such as *Melosira sphaerica* at Signy Island, while pools and rocks receiving enrichment from bird colonies have strands of *Enteromorpha bulbosa*.

A wide range of microalgae live on or within the larger plants (macroalgae) and animals of the benthos. For example, diatom webs of *Biddulphia punctata* and *Nitzschia sublineata* have been observed covering the brown seaweed *Desmarestia*. Films of diatoms cover the sponge spicule mats (Palmisano *et al.*, 1985c), molluscs (Mullineaux & De Laca, 1984) and other benthic fauna in McMurdo Sound.

Overwintering populations of algae within the sediments may provide the inoculum for sea-ice and planktonic communities, as well as a major source of carbon for planktonic and benthic heterotrophs during winter when there is no light available for photosynthesis. In McMurdo Sound, for example, chlorophyll *a* could be measured in the sediments throughout the year (Fig. 6.3), and large amounts of sediment were collected in traps

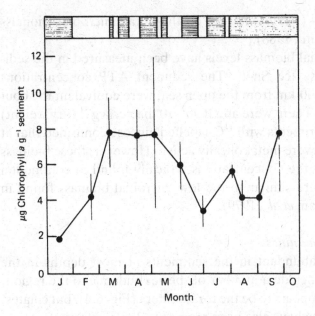

Fig. 6.3. Chlorophyll *a* in the surface 2 cm of the sediments in McMurdo Sound, offshore from McMurdo Station. The shading at the top indicates periods of ice cover. Modified from Berkman *et al.* (1986).

during a period of water column mixing in winter when intense winds caused a breakout of sea ice (Berkman, Marks and Shreve, 1986). Winter samples of these surficial sediments contained viable algae that grew when exposed to artificial light.

### 6.3.2 *Bacterial populations*

Bacterial populations in antarctic sediments can achieve extremely high biomass levels. Phospholipid analysis of sediments off Cape Armitage, Ross Island, revealed concentrations of microbial biomass equivalent to a subtropical Florida estuary. However, the productivity of this antarctic assemblage is low and these benthic microbial communities are probably very old (White, Smith & Stanton, 1984).

A detailed analysis of the fatty acid composition of the sediment microbiota in McMurdo Sound has also revealed the presence of anaerobic bacteria, specifically the sulphate-reducing *Desulfovibrio* and *Desulfobacter*. These organisms presumably live in reduced microniches in the oxidised marine sediments. The fatty acid analysis suggested that both bacteria were present in significant numbers, accounting for 1–2% (*Desul-*

*fovibrio*) and 5–10% (*Desulfobacter*) of the total microbial biomass (Smith, Nichols & White, 1988).

Much lower microbial biomass levels have been measured in the sediments under the Ross Ice Shelf. The sediment ATP concentrations beneath the J9-hole, 400 km from the open sea, were equivalent to about 1 mg bacterial C m$^{-2}$. There were about $10^7$–$10^8$ bacteria g$^{-1}$ (dry weight) of sediment, and experiments with $^{14}$C-labelled glucose confirmed that at least some of the cells were metabolically active. However, these biomass values are well below the concentrations typically found in continental shelf sediments, and are similar to the low microbial biomass found in abyssal sediments (Azam *et al.*, 1979).

### 6.3.3   *Protozoan communities*

Protozoa are abundant in the sediments at most depths in the Southern Ocean and include a number of species endemic to the region. The dominant group appears to be the foraminifera (Fig. 6.4), but ciliates, flagellates and amoebae have also been reported.

A remarkable tree-shaped foraminifer has been discovered in the McMurdo Sound benthos (De Laca, Lipps & Hessler, 1980). This large species, named *Notodendrodes antarctikos*, stands up to 38 mm high (average 20 mm) and the average protoplast has a mass of about 1 mg wet weight. The test comprises a stem that bifurcates into a crown of branches, anchored into the sediments with a double-walled bulb and dendritic root system. Cytoplasm extends throughout this complex test and penetrates through the walls of the branches and stem to form clusters of pseudopodia which extend into the water. These pseudopodia capture particulate material which is digested outside the test. The cytoplasm does not penetrate through the root sheath, but the root system appears to be an important site for the uptake of dissolved organic compounds (see Section 6.4.3). This species occurred on the silt, sand and mud sediments on the western side of McMurdo Sound at population levels of about 100 cells m$^{-2}$ (De Laca, 1982a).

Foraminifera species, including *Notodendrodes*, demonstrate a clumped distribution over the sediments. The occurrence of some species appears to be affected by the abundance of the macrofauna which provide a stable physical substrate for colonisation. Two calcareous foraminifers in western McMurdo Sound, *Cibicides colbecki* and *Rosalina globularis*, occurred more abundantly on the shells of the bivalve *Adamussium colbecki* than on the adjacent sediments, while two species of *Trochammina* were more abundant in the sediment. The larger shells (>43 cm$^2$) supported a much

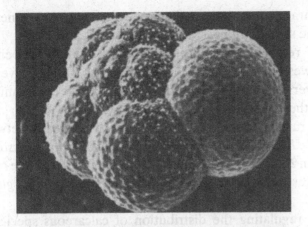

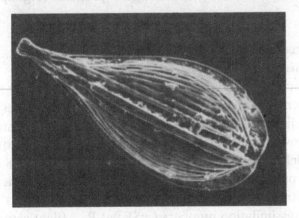

Fig. 6.4. Foraminifera from the Ross Sea. Top, *Neogloboquadrina pachyderma*; middle, arenaceous species with a test of sponge spicules; bottom, *Fissurina texta*. Scanning electron micrographs by Dr Barbara L. Ward, Geology Department, Victoria University of Wellington, New Zealand. Reproduced from Ward (1984) by permission of the author.

higher density of foraminifers than smaller ones. A low density zone was observed around the shell margins. (Mullineaux & De Laca, 1984).

There are large regional differences in the distribution of benthic foraminifera. The western side of McMurdo Sound has a higher diversity of species and a lower total abundance and biomass than the community at Cape Armitage on the eastern side of the Sound (De Laca, 1982b).

The bathymetric range of antarctic foraminifera varies greatly between species. For example *Bathysiphon filiformis* and *Rhabdammina abyssorum* range to greater than 2000 m depth (Milam & Anderson, 1981) while other species seem to be restricted to shallow habitats e.g. *Trochammina glabra* (Ward, 1984).

A critical factor regulating the distribution of calcareous species of foraminifera is the calcium carbonate compensation depth (CCD). The solubility of calcium carbonate increases with increasing salinity and pressure and below a certain depth such species are unable to maintain their calcite test structure. In much of the world's oceans the CCD is very deep, commonly below 4000 m, but the cold waters of the Southern Ocean hold high concentrations of carbon dioxide which favours calcite dissolution, and the CCD can be at much shallower depths. In McMurdo Sound this solution boundary seems to be at about 620 m in open water and only 230 m in harbours and enclosed basins (Ward, 1984). These appear to be the lower depth limits for virtually all calcareous species, and at greater depths the assemblage shifts towards arenaceous species, i.e. those that produce a test of cemented sand grains or other particles (e.g. sponge spicules, Fig. 6.4).

## 6.4    Microbial processes
### 6.4.1    *Photosynthesis*

Although microalgae have been observed at many shallow-water sites there have been few measurements of benthic photosynthesis. A population of the diatom *Trachyneis aspera* was observed forming a flocculant golden-brown layer over sponge spicule mats 20–30 m below the annual sea ice in McMurdo Sound (Palmisano *et al.*, 1985c). This large (100–400 $\mu$m long) pennate diatom showed several features indicating adaptation to the extremely dim light environment ($<0.6$ $\mu$E m$^{-2}$ s$^{-1}$) in which it grows. The assimilation numbers ($\pm$SE) at P$_{max}$ (light-saturated photosynthesis) were $0.21 \pm 0.03$ mg C (mg chl$a$)$^{-1}$ h$^{-1}$, photosynthetic efficiency ($\alpha$) was $0.022 \pm 0.005$ mg C (mg chl$a$)$^{-1}$ h$^{-1}$ ($\mu$E m$^{-2}$ s$^{-1}$)$^{-1}$ and the irradiance at saturation (I$_k$) was $11.2 \pm 1.0$ $\mu$E m$^{-2}$ s$^{-1}$. Unlike many shade-adapted species, however, this diatom population was not photoinhibited at irradiances up to 300 $\mu$E m$^{-2}$ s$^{-1}$, well above I$_k$. This species is not a component of the sea-ice community, but seems to be restricted to the benthos, particularly on the siliceous spicule surfaces and in the interstitial water between the sponge spicules.

*In situ* photosynthetic measurements of the benthos in McMurdo Sound revealed extremely low microalgal photosynthetic rates, for example 51–95 mg C m$^{-2}$ h$^{-1}$ (0.06–0.1 mg C (mg chl$a$)$^{-1}$ h$^{-1}$) at 18 m depth off Cape Evans (Dayton *et al.*, 1986). The slowest rates were recorded on the eastern side of the Sound – 0.1–2.6 mg C m$^{-2}$ h$^{-1}$ (0.003–0.02 mg C (mg chl$a$)$^{-1}$ h$^{-1}$). These slow and often undetectable levels of photosynthesis suggest that much of the chlorophyll in the sediments is associated with inactive, but undecomposed cells.

### 6.4.2  *Bacterial heterotrophy*

Rates of bacterial metabolism in the antarctic sediments appear to be extremely slow, possibly limited by temperature and the low inputs of organic carbon through most of the year. Despite a high microbial biomass in the McMurdo Sound sediments the rate of bacterial DNA synthesis was some 300-fold slower than in estuarine sediments at lower latitudes. The average growth rate measured in McMurdo Sound with tritiated thymidine was 7.6 × 10$^4$ cell divisions h$^{-1}$ g$^{-1}$ of sediment (Smith, Coulson & Morris, 1986). The bacterial growth rates in a non-spicule sponge mat increased by a factor of three with the input of sea-ice algae during melt-out in late summer (White *et al.*, 1984).

Evidence that factors other than temperature influence sediment metabolism has come from studies near Kerguelen Island (Delille & Cahet, 1985). In contrast with the oceanic region south of the Polar Front, the water temperatures around this subantarctic island vary greatly with season, from about 10 °C in December–January to 1–2 °C in August–September. However, the maximum heterotrophic potential in the sediments occurred during autumn, and the maximum bacterial counts were measured in winter when temperatures were at their annual minimum. Presumably this pattern is dictated by the timing of allochthonous organic inputs to the system from terrigenous as well as marine sources such as the kelp beds.

Aerobic degradation of the antarctic kelp-like species *Himantothallus grandifolius* indicated that macroalgal detritus is formed by bacterial action as efficiently in antarctic coastal waters as in the temperate zone (Reichardt & Dieckmann, 1985). Fresh debris of this brown alga was initially dominated by small, mostly coccoid bacterial cells in the size range

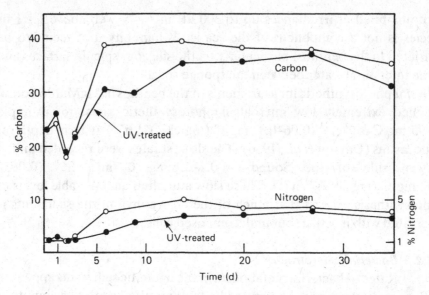

Fig. 6.5. Carbon and nitrogen accumulation in antarctic macroalgal detritus during its aerobic decomposition. The pretreatment with ultraviolet radiation slowed this accumulation, suggesting that it was due to microbial colonisation and growth. Modified from Reichardt & Dieckmann (1985).

$0.1$–$0.4$ $\mu$m. Colonisation was mainly along the junctions of epidermal cell walls, a feature also observed in decomposing kelp at temperate latitudes. After a lag of about three days bacterial growth rates increased and large rods became the dominant bacterial forms. During this exponential growth phase the apparent doubling time was 27 h and bacterial densities rose to about $4 \times 10^5$ cells $mm^{-2}$ of tissue, comparable with decomposing marine algae at lower latitudes. Over this period of growth the particulate carbon content of the detrital particles doubled and the particulate-N levels increased four-fold indicating that the microbial flora was scavenging compounds from the surrounding water. An initial pretreatment of the detritus with ultraviolet radiation slowed the rate of C and N accumulation supporting the conclusion that at least some of the observed changes were due to microbial growth (Fig. 6.5). This partially decomposed but C- and N-rich detritus appeared to be preferred by intertidal amphipods as a food source over fresh algal material.

### 6.4.3 *Protozoan physiology*

At least some foraminifers in the antarctic benthic environment meet an important fraction of their energy requirements through uptake of dissolved organic compounds. *Notodendrodes antarctikos* was capable of transporting a wide range of dissolved free amino acids at ambient substrate levels in McMurdo Sound. Uptake rates were relatively constant over the temperature range $-1.5-10\,°C$. Glutamic acid uptake was saturated at about 50 $\mu$M with a half saturation constant ($K_t$) of 6.85 $\mu$M. Respiration rates for this species averaged ($\pm$SD) $134 \pm 40\ \mu g\,O_2\,g^{-1}\,h^{-1}$ at $-1\,°C$ and increased with increasing concentrations of protein hydrolysate. The measured uptake velocities of amino acids were calculated to satisfy 86% of this respiratory demand for carbon (De Laca, 1982a).

## 6.5    Trophic structure

The marine benthos on the antarctic continental shelf is rich in species and abundance, and the food web is complex. Detrital material from seaweed beds, sea-ice algae and phytoplankton provide the main inputs of carbon and energy, although the particles may be considerably modified by attached bacteria before they reach the sea floor. Among the animals, sponges are often the biomass dominants and are sometimes limited more by predation than by food supply. In parts of McMurdo Sound, for example, sponges cover 55% of the sediments and provide important vertical structure to the community (Dayton *et al.*, 1974). The relatively fast-growing sponge *Mycale acerata* can grow over and smother other sessile animals but the spread of this species is severely restricted by predation by two starfishes, *Perknaster fuscus antarcticus* and *Acodontaster conspicuus* and by the nudibranch *Austrodoris mcmurdensis*. These predators are in turn kept in check by another starfish, *Odontaster validus*. More than 20 species of sponge inhabit McMurdo Sound and provide a refuge to the fish from seal predation, as well as elevated perches for animals such as fish, holothurians, crinoids and ctenophores. Hydroids and actinians provide food to aeolids and pycnogonids (sea spiders). Detritus feeders include the urchin *Sterechinus neumayeri* and the starfish *Odontaster*. The actinian *Urticinopsis antarcticus* consumes echinoderms and medusae that brush against it (Dayton *et al.*, 1974). Invertebrates inside the sponge mats include giant nematodes of the genus *Deontostoma*.

Foraminifera may provide an important link between the bacteria or even dissolved organic compounds associated with decomposing phytoplankton cells, and larger animals. However, sedimented diatoms may directly enter the diets of certain macrofaunal species. For example, the stomach contents of the *Ophionotus victoriae*, one of the most widely distributed and abundant brittle stars on the antarctic continental shelf, included 52 species of diatoms (Kellogg *et al.*, 1982). A major sterol found in the sea-ice diatom *Nitzschia cylindrus* has also been identified in the stomach contents of *Odontaster* and *Deontostoma* at a benthic Cape Armitage site. These stomach content samples additionally contained significant levels of three fatty acids that are found within sea-ice diatoms (Smith, White & Nichols, 1986) further supporting a direct transfer pathway of carbon and energy from microalgae to large animals.

# 7

# Lakes and streams

## 7.1   Introduction

Lakes and pools provide some of the most favourable habitats for microbial growth in the antarctic region. Many are capped by thick ice and the water beneath remains unfrozen despite winter air temperatures well below 0 °C. The smaller pools freeze solid each year, but melt out in summer to provide relatively warm and often nutrient-rich conditions for the growth of microalgae and associated heterotrophs. Most of these aquatic environments lie near the coast and are influenced by the sea through marine aerosols, seaspray or enrichment by marine animals. Some still retain a direct connection with the sea, while others are derived from trapped seawater that has been further concentrated by evaporation and freezing. Unlike the marine ecosystems, however, inland antarctic waters contain very few species of multicellular plants and animals and the community structure in some is entirely microbial.

Flowing water habitats are also a feature of the coastal Antarctic. Most are small seeps or streams with discharges of only a few litres per second, but rivers with discharge peaks of several cubic metres per second are known from various sites around the continent. All of these waters are fed by melting snowbanks or glaciers and generally flow for only the warmest weeks or months each summer. Some are highly turbid and contain sparse microbial communities, probably because of the abrasive effects of their suspended sediment load. The clearer streams, or those with slow-moving water, often contain rich epilithic mats and films most commonly dominated by cyanobacteria or chlorophytes.

## 7.2   The environment

Most antarctic lakes and streams are covered by ice through all but a few weeks of the year. This single feature has a pervasive influence on their physical, chemical and biological properties, and sharply distin-

Fig. 7.1. A permanent ice cap 5 m-thick with numerous pinnacles overlies Lake Hoare in the Dry Valleys region.

guishes them from inland waters at non-polar latitudes. In some regions of the continent this thick ice cap overlying the lake persists even during summer (Fig. 7.1). Freshwater ice lacks the brine channels and open platelet structure which provide a microbial habitat in sea ice, but it fundamentally controls the hydrodynamics, light regime and other environmental variables which regulate the planktonic and benthic micro-biota. Unusual salinities and temperatures are also a feature of many antarctic waters, particularly the ice-covered meromictic lakes of the continent. Discharge and silt load are two additional environmental variables that appear to regulate the distribution and activity of stream microbiota, and are therefore also examined here.

### 7.2.1  Snow and ice cover

The thickness of the ice cover depends on three processes: the transmission and absorption of sunlight, the conduction of energy out of the ice to the surrounding water and ice, and the release of the latent heat of fusion at the ice–water interface. Under steady state conditions the release of latent heat at the bottom-ice surface by freezing is controlled by the rate of ablation from the upper surface. McKay *et al.* (1985) derive the following relationship for thick lake ice:

$$Z = (k\Delta T - S(1 - a)(1 - r)h)/L \qquad (7.1)$$

where  $Z$ = the equilibrium thickness of the ice cover

$k$ = thermal conductivity of the ice

$\Delta T$ = difference between the temperature at the ice–water interface (generally 0 °C) and the average annual air temperature immediately above the ice.

$a$ = albedo of the ice

$r$ = fraction of the ice covered by dark absorbing material such as sand and silt

$h$ = extinction pathlength of the ice (cosine of the solar zenith angle divided by the attenuation coefficient of the ice)

$S$ = solar radiation incident on the lake ice

$L$ = the heat flux due to latent heat released at the ice–water interface, given by the relationship:

$$L = v\varrho c_f \qquad (7.2)$$

where  $v$ = rate of formation of new ice, $\varrho$ = the density of ice, and $c_f$ = latent heat of fusion.

For reasonable estimates of these parameters for the Dry Valley lakes (*c.* 78 °S, 163 °E, see Appendix 1) and a measured ablation rate of 30 cm $y^{-1}$ this model predicts ice cover from about 3.5 m (annual air temperature = −20 °C, $S = 104$ W $m^{-2}$, $r = 0.1$, $h = 1.03$) to 5.5 m (same, but $r = 0.25$), which compares favourably with observed ice thicknesses of 3–6 m. One Dry Valley lake, Lake Vida, is permanently frozen to its base at 11 m. This unusually thick ice is presumably allowed by the minimal refreezing at its sediment–ice interface, i.e. the denominator in equation (7.1) approaches zero and therefore $Z$ becomes large (McKay *et al.*, 1985).

The penetration of photosynthetically available radiation (PAR) through the ice-cap to the phototrophic populations below is similarly dependent upon the albedo ($a$) and the fraction of the ice covered by dark absorbing material ($r$), the attenuation coefficient for the ice ($k_i$), and also the ice thickness ($Z_i$):

$$I' = (1 - a)(1 - r)I_o \exp(-k_i Z_i) \qquad (7.3)$$

where $I'$ is the PAR immediately under the ice-cap, and $I_o$ is the incident PAR, typically about one-half of $S$ in equation (7.1).

This large number of independent variables can result in marked differences in the underwater light field between nearby lakes, or over relatively short time or length scales.

The albedo term $(1 - a)$ is especially important at polar latitudes because of the continuously low solar angles and the high reflectance of snow cover (see Appendix 2). In Signy Island lakes, for example, the loss of the 10–30 cm layer of snow in September–October increases the penetration of PAR through the 1 m-thick ice from <1% to >20% (Hawes, 1985). Since $a$ is so large at polar latitudes, small changes in solar angle with time of day can have a major influence on $(1 - a)$, and sub-ice populations may experience a greater diurnal variation in the PAR field than measurements of above-ice incident PAR would suggest.

Changes in $r$ or $a$ across the lake ice surface may cause a much greater degree of spatial heterogeneity in the underwater light field than is experienced in ice-free lakes. On certain Dry Valley lakes the distribution of wind-blown sediments and moraine material ($r$) can cause major variations in ice thickness and light penetration. These effects are spectacularly illustrated on Trough Lake and Lake Miers where ice mounds capped by glacial moraine including boulders stand up to 3 m above the mean lake ice level (Fig. 7.2).

Local wind patterns can strongly influence the distribution of snow, and thus changes in albedo ($a$). An empirical model for a subarctic lake showed that the penetration of light through the ice was inversely related to snow cover and the distribution of snow-derived white ice (Roulet & Adams, 1984). The transmission of incident PAR varied from less than 0.02% in the marginal and downwind areas to 4.5% in the exposed central area of the lake.

The absorption and scattering of light by the lake ice markedly reduces penetration of PAR to the water column beneath. The value of the attenuation coefficient ($k_i$, equation (7.3) above) varies greatly with the type of ice. For two Dry Valley lakes, incident PAR was reduced to 14% beneath the relatively clear 3.25 m ice cap of Lake Vanda but to 0.9% beneath 4.5 m of more bubbly ice over Lake Fryxell.

In the water column beneath the ice the average PAR experienced by the total phytoplankton community ($I_{av}$) is given by:

$$I_{av} = (I'/(k_w \cdot \Delta z))(1 - \exp(-k_w \Delta z)) \tag{7.4}$$

where $k_w$ is the average attenuation coefficient for the water column from immediately beneath the ice cap to the average depth limit of the phytoplankton ($\Delta z$).

Fig. 7.2. Glacial moraine including gravels and boulders are distributed across the 6 m-thick ice cap of Trough Lake, southern Dry Valleys. Reproduced from Howard-Williams, Vincent & Wratt (1986b) by permission of New Zealand Antarctic Record.

In practice the relatively unmixed waters beneath the ice-cover may allow different communities to adjust to PAR at specific depths, and a measure of $I_{av}$ may be of less interest than PAR experienced by the phototrophs at a specific depth, $z_p$:

$$I_z = I_o(1 - a)(1 - r) \exp (-k_i z_i - k_w \Delta z_p) \tag{7.5}$$

Even within the water column, however, the attenuation coefficient ($k_w$) may change markedly with depth, and equation (7.5) must be evaluated stratum by stratum.

Light quality is also affected by lake ice, although to a much lesser extent than under the algal-rich sea ice (see Chapter 3). The spectral distribution of PAR under the ice resembles that for distilled water with a substantial reduction of red light and maximum transmission in the blue and green wavebands (Fig. 7.3).

Lake ice greatly reduces the transfer of wind energy and momentum to the water column beneath, and many ice-capped lakes in Antarctica experience physically stable conditions throughout the year. The apparent absence of wind-induced turbulence favours certain microbial groups such as flagellates and buoyant cyanobacteria which can adjust their position in

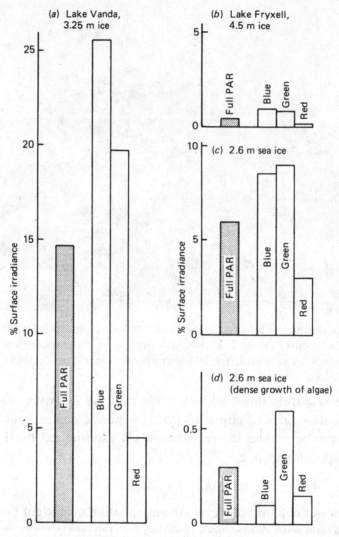

Fig. 7.3. The influence of ice cover on the spectral composition of photosynthetically available radiation (PAR). The Lake Vanda and Lake Fryxell (Dry Valleys) data are from original observations by W. F. Vincent and G. S. Wratt; the sea-ice data are from Palmisano *et al.* (1986); note the ten-fold expansion of vertical scale in (d). The values are for full PAR or separate wavebands, immediately beneath the ice.

the water column. Small cells have both slow sinking velocities (sinking is proportional to the square of the radius of a particle) and a high light-capturing ability per unit pigment (Raven 1984), and are therefore especially favoured in the dim-light, quiescent environment beneath the ice.

In the deep strata of some lakes the dominant transfer mechanism may be molecular diffusion and sedimentation which are much more readily defined than turbulent diffusion. This has greatly simplified, for example, the formulation of biogeochemical models for Lake Vanda (e.g. Canfield & Green, 1985). These slow rates of transfer allow certain microbial end-products and intermediates to accumulate to high concentrations in specific depth strata (see Section 7.4.3).

Ice formation may also have a direct influence on the chemical environment in antarctic waters. As the water freezes some of the dissolved constituents are excluded from the ice and are forced back into the underlying water column. In Heywood Lake (Signy Island) dissolved organic carbon concentrations rose to 9.3 mg $l^{-1}$ in the surface waters during winter freezing, while bottom water concentrations remained at less than half of this value (Ellis-Evans, 1981a). In Ace Lake (Vestfold Hills) the melting of the fresher ice in summer, in combination with meltwater inputs, resulted in a surface layer of low salinity water overlying a weak chemocline at 3 m. Dissolved gases are also expelled during the freezing process, and in Ace Lake dissolved $N_2$ beneath the ice was in greater concentration than for water of the same salinity in equilibrium with the atmosphere (Hand & Burton, 1981). This effect on dissolved $N_2$ has been described in detail for Lake Hoare (Fig. 7.1) in the Dry Valleys (Wharton *et al.*, 1987). By assuming that the microbiological processes consuming (e.g. $N_2$-fixation) or producing (e.g. denitrification) nitrogen gas were negligible relative to this abiotic concentration mechanism it was possible to use the ratio of $N_2 : O_2$ dissolved in the water to estimate the relative importance of biological processes on the oxygen budget for this lake. These calculations showed that 58% of the net oxygen production in Lake Hoare could be accounted for by freeze-concentration of the inflowing meltwater, while the remaining 42% was attributable to net photosynthetic production.

The elevated concentrations of dissolved gases within ice-covered lakes are additionally favoured by very slow rates of diffusion through the ice cap, and by slow rates of horizontal mixing between water in the main body of the lake and water in the edge zone which may come in contact with the atmosphere if the edge of the ice cap melts out in summer. This peripheral 'moat' is a feature of most of the Dry Valley lakes, and is probably the major site of gaseous efflux from them (Wharton *et al.*, 1987).

In the Dry Valley lakes oxygen levels are typically 200–400% of air-equilibrium values (e.g. Parker *et al.*, 1982b), and in several of the lakes deep strata of nitrifying bacteria have produced peaks of nitrous oxide from 500 to 30 000% of air-equilibrium concentrations (e.g. Vincent, Downes & Vincent, 1981). It is not clear to what extent these conditions of perennial supersaturation have an effect on the microbiota, but it is possible that the high oxygen tensions may directly influence some microbial processes, e.g. stimulation of oxygenase activity (photorespiration) relative to carboxylase (photosynthesis) activity of the enzyme ribulose-1,5-bisphosphate carboxylase-oxygenase (see Section 7.4.2).

Although lake ice inhibits mixing a number of hydrodynamic processes may continue to operate, and the assumption of molecular diffusivities may not always be justified. In permanently ice-covered Lake Vanda, for example, thermohaline convective mixing occurs in specific layers of the upper water column. Radioisotope releases in one of the convection cells have provided clear evidence of strong horizontal currents (*c.* 1 cm s$^{-1}$) accompanied by vertical turbulent diffusion (Ragotzkie & Likens, 1964).

Currents entering the lake as buoyant inflows may provide an important vertical exchange mechanism for some ice-covered waters. For example Heywood Lake on Signy Island receives a substantial input of water from cold, meltwater streams. Fresh water has its maximum density at 3.98 °C, and above and below this temperature its density ($\varrho_w$, in kg m$^{-3}$) may be estimated as:

$$\varrho_w = -\alpha 1000 \, (T - 3.98)^2 + 1000 \tag{7.6}$$

where $\alpha = 8 \times 10^{-6}$ (°C$^{-2}$), and

   $T$ is temperature (°C)

This algorithm approximates $\varrho_w$ to within 4% over the range 0–8 °C (Matthews & Heaney, 1987). The inflows to Heywood Lake are colder and thus less dense than the 2.5 °C water column. This incoming water therefore flows through the lake as a discrete, low density layer immediately beneath the ice-cap. At the estimated current velocities (0.5–1.0 cm s$^{-1}$) this throughflow has a Richardson Number close to the threshold of transition from laminar to turbulent flow, and may generate turbulent mixing that propagates down through the water column.

A wide range of density effects may generate convective mixing beneath the ice in cold waterbodies such as Heywood Lake. In the shallow littoral zone the water warms towards 3.98 °C much faster than in the middle of the lake, and thus increases in density. This water can then flow down the side of the lake basin as a density current. Saline groundwaters, or stream-waters entering the lake with temperatures closer to 3.98 °C than the lakewater, could generate similar currents that sink to their depth of neutral buoyancy, entraining water as they sink. If the cold lakewater has a high attenuation coefficient then incoming radiation may warm the water immediately under the ice, producing convective plumes that penetrate down into the water column beneath. Ice exclusion of salts during freezing generates convective mixing beneath ice shelves and sea ice in the Southern Ocean (see Chapter 3), and similar density plumes might occur during the freeze-up of antarctic lakes. None of these density effects is well defined in antarctic lakes, but observations at Heywood Lake (Hawes, 1983c) pro-vide strong evidence that mixing processes operate beneath thick ice cover. Theoretical calculations by Matthews & Heaney (1987) indicate that the vertical mixing velocities generated by penetrative convection in Heywood Lake could be up to 2 m d$^{-1}$.

### 7.2.2 Salinity

Many of the inland waters of Antarctica are derived from the sea and therefore contain high concentrations of dissolved solids. Other sources of the salts include marine aerosols and the geochemical weather-ing of catchment soils and rock, but the relative importance of these various inputs is often unclear. Volcanic and geothermal inputs to antarctic lakes are rarely important.

Some lakes contain seawater that has been concentrated by freezing and evaporation. When seawater freezes calcium carbonate is the first salt to precipitate out, then sodium sulphate (at −8.2 °C), sodium chloride (−22.9 °C), and magnesium and potassium chlorides (−36 °C). This freez-ing process ultimately produces a calcium chloride brine (Thompson & Nelson, 1956). Lakewaters that contain high concentrations of these salts, or surface salt deposits in their catchments are known from throughout the antarctic region. Thick beds of mirabilite ($Na_2SO_4 \cdot 10H_2O$) occur on the McMurdo Ice Shelf (Brady, 1980) and encrust the rocks surrounding various lakes in the Vestfold Hills. The sediments of Lake Bonney (Dry Valleys) contain layers of halite (NaCl), hydrohalite (NaCl $\cdot$ 2H$_2$O) arago-nite ($CaCO_3$) and gypsum ($CaSO_4 \cdot 2H_2O$). The dominant salt in hyper-saline Don Juan pond (Dry Valleys) is calcium chloride which precipitates out as antarctite ($CaCl_2 \cdot 6H_2O$), a mineral peculiar to this basin (Torii & Ossaka, 1965).

The concentration and composition of the lake salts have important implications for microbial distribution and activity. The total dissolved solids content of Don Juan Pond can exceed 500 kg m$^{-3}$, and although several bacterial species and a yeast have been isolated from this habitat it is unlikely that any organism can metabolise and grow at this osmolarity. Deep Lake (Vestfold Hills) is also hypersaline, but much less so (270 kg m$^{-3}$) than Don Juan Pond, and contains populations of the halophilic green alga *Dunaliella*. High sulphate levels derived from the sea provide an important electron acceptor for some bacteria. In Ace lake (Vestfold Hills), sulphate-reducing bacteria have completely consumed the marine-derived $SO_4$ in the bottom anoxic zone.

Salinity also exerts a wide range of indirect effects on antarctic microbiota by influencing their physical environment, or altering aspects of their chemical environment in addition to osmolarity. Saline waters remain liquid at temperatures well below 0 °C. The relationship between the temperature of freezing ($T_f$) and salinity ($S$, in ‰) may be approximated by:

$$T_f = -0.003 - 0.0527S - 0.00004S^2 \tag{7.7}$$

This algorithm gives close estimates up to and near seawater salinities (Maykut, 1985), but may substantially underestimate actual $T_f$ under hypersaline conditions. For example it predicts a $T_f$ for Deep Lake water of −7 °C but the actual freezing point has been measured at −28 °C (Kerry *et al.*, 1977).

Dissolved salts not only increase the density of water ($\varrho_w$) but also increase the change in density for a given change in temperature. For example, if equal volumes of freshwater and Deep Lake water were warmed from 4 °C to 9 °C, both waters would decrease in density but the change in $\varrho_w$ would be eleven times greater for the Deep Lake water. This influence of salinity on the temperature–density relationship means that saline lakes may be more prone to density stratification for a relatively small heat input or thermal gradient (Ferris & Burton, 1987).

Salinity also affects the temperature of the water at which its maximum density is attained ($T_m$) which in turn influences the mixing regime as described above. $T_m$ drops linearly with increasing salinity over the range 0–27‰ by the relationship:

$$T_m = 3.98 - 0.216\,S \tag{7.8}$$

Above 27‰, maximum density is attained at the temperature of freezing ($T_m = T_f$; Maykut, 1985).

Saline waters also contain much lower concentrations of dissolved gases when in equilibrium with the atmosphere. For dissolved oxygen, nitrogen and argon the saturation or air-equilibrium concentration may be calculated from Weiss's (1970) algorithms.

### 7.2.3   Temperature

Ice cover and salinity are important determinants of the thermal regime in antarctic lakes. The formation of ice insulates the water column from low ambient air temperatures in winter, and additionally releases large amounts of heat during he freezing process by the latent heat of fusion (see Appendix 2). The buffering effects of ice formation on seasonal temperature are clearly seen by comparing ice-covered Ace Lake with nearby ice-free Deep Lake in the Vestfold Hills (Fig. 7.4).

Waters that are stabilised by salt gradients can absorb solar radiation and rise to temperatures well above even the summer air maxima. The most dramatic example known from the Antarctic is Lake Vanda in which temperatures rise from near 0 °C immediately beneath the permanent ice

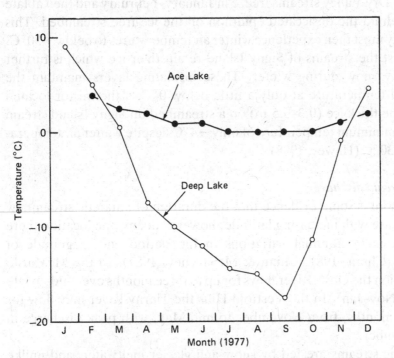

Fig. 7.4. Water temperatures at 3 m depth in two nearby lakes in the Vestfold Hills. Ace Lake was covered by 1–2 m of ice from late April to December; data from Hand & Burton (1981). Deep Lake had no ice cover throughout 1977; data from Ferris & Burton (1987).

cap to 24 °C at the bottom of the lake. This inverse distribution of heat is stabilised by the increasing salinity with depth, from about 0.1‰ near the surface to 123‰ at the bottom.

Hypersaline waters on the antarctic continent have theoretical freezing points well below 0 °C (see Section 7.2.2) and become extremely cold during winter. The surface waters of Deep Lake, for example, cool to −18 °C in winter (Fig. 7.4) and the water column remains ice-free, isothermal and free-mixing over the period July–September.

Stream temperatures in Antarctica generally lie in the range 0–5 °C, but daily maxima above 10 °C are known from even the southernmost flowing waters. Diel changes in water temperature can be substantial, with freezing to a variable extent from 0100 to 0700 h and maximum temperatures in the late afternoon, 1400–1600 h. For streamwaters flowing into lakes these diurnal variations in temperature cause variations in density, and therefore the depth to which the inflow sinks.

Snow and ice cover can effectively insulate the streambed communities from large fluctuations in temperature. As an extreme example of this effect, the Dry Valley streams freeze in January–February and then ablate leaving behind the desiccated epilithon on the ice-free streambed. This community must then experience winter air temperatures to below −50 °C. By contrast the streams of Signy Island retain their ice which is further covered by snow during winter. These insulating layers maintain the streambed temperature at only a little below 0 °C: a thermistor located beneath the thick ice (0.3–0.5 m) on a streambed on Signy Island stream showed a minimum temperature of only −4 °C despite winter air temperatures to −30 °C (Hawes, 1988).

### 7.2.4    Stream discharge

There is some evidence that the duration of antarctic streamflow may decrease with increasing latitude, however in any one location there can be large interannual variations in the period and magnitude of discharge (Chinn, 1981; Chinn & McSaveney, 1987). In the McMurdo Sound region the Onyx River flows for up to three months over mid- to late summer (Nov–Jan). In the Vestfold Hills the Tierny River may flow for about six months, from November to mid-May, with peak discharge in mid-December.

Antarctic streams are fed by snow and glacier meltwater, and unlike temperate regions where discharge is closely linked to incoming precipitation, their variations in flow are controlled primarily by solar radiation. There is a strong positive correlation between mean daily discharge and the seasonal radiation cycle, although this pattern may be modified in higher order streams that intermittently receive water from tributaries and lake overflows. Solar radiation continues to influence discharge over the 24 h cycle, and many streams show pronounced diel variations that depend upon the position of the sun relative to the melting glacier face. In Northern Rookery Stream, for example, the discharge was found to vary from $2\ 1\ s^{-1}$ at noon when the glacier face was in shadow to more than $100\ 1\ s^{-1}$ when exposed to direct sunshine at 1500 hours (Fig. 7.5). Diel formation and melting of ice dams in the larger streams may further modify short-term changes in discharge.

These high amplitude fluctuations are much reduced in streams that flow from lakes. For example Trough Lake at the head of the Alph River has sufficient storage to damp the diel melting effects; its discharge varies by less than 5% over the course of 24 h (Howard-Williams, Vincent & Wratt, 1986b). Similarly, the diel hydrograph of Tierny River at Ellis Rapids (Vestfold Hills) is dampened by storage in the upstream Friendship Lakes system (Colbeck, 1977).

Snow is a much more important source of water for streamflow in the maritime Antarctic than on the continent where glaciers are typically the primary meltwater source. On Signy Island melting snowbanks accumulated over winter produce a spring peak in discharge (Hawes, 1988). By contrast, the Dry Valleys receive very little snowfall (see Appendix 1) and the production of streamwater from melting glacier ice often rises with increasing air temperatures during summer. A much more complex seasonal hydrograph, however, characterises the larger river systems with

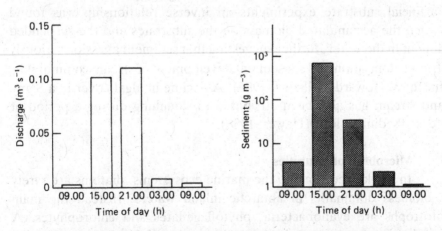

Fig. 7.5. Diurnal variations in discharge and sediment content of Northern Rookery Stream, Cape Bird. Data from Howard-Williams *et al.* (1986a).

variable storage components including ice-dammed lakes (e.g. the Onyx River).

### 7.2.5    *Stream sediment load*

High silt loads are a characteristic feature of glacial streams in temperate as well as polar zones, and appear to have a major influence on the development of stream microbiota. Streams of the southern Victoria Land region carry a wide range of silt loads (Howard-Williams *et al.*, 1986a) from the highly turbid Commonwealth Stream (1016 g sediment $m^{-3}$) which cuts steeply through fine, unconsolidated sediments at the entrance to the Taylor Valley, to the clearwater (0.9 g $m^{-3}$) outflow stream from Lake Miers, which presumably acts as a settling basin for the inflowing sediment. The sediment content of glacial ice can be high (e.g. 101 g $m^{-3}$ in icicles on the glacier feeding Fryxell Stream, Taylor Valley) but much of this may be trapped in the moraines near the glacier face, and downstream turbidity can be relatively low (e.g. 2 g $m^{-3}$ in the lower reaches of Fryxell Stream). The sediment load can vary markedly with discharge, and thus with time of day. In Northern Rookery Stream, for example, there was a sudden rise each day in the silt load corresponding with the abrupt diurnal increase in discharge (Fig. 7.5).

Sediment load appears to be a major variable controlling the distribution of stream microbial mat communities. Clearwater streams (<10 g sediment $m^{-3}$) in the McMurdo Sound region have well-developed epilithic mats and films, while in turbid streams these communities are restricted to slow-moving side arms or wide areas of multiple channels where flow velocities and associated sloughing effects are greatly reduced. In a series of artificial substrate experiments an inverse relationship was found between the accumulated biomass on the substrates and the suspended sediment in the water, further suggesting that sediment abrasion seriously restricts colonisation and subsequent development of stream communities (Vincent & Howard-Williams, 1986). A decline in algal cover in a Signy Island stream has also been attributed to sloughing during a period of increased sediment load (Hawes, 1988).

## 7.3    Microbial communities

In marked contrast to the marine ecosystems, diatoms are rarely the biomass dominants in antarctic inland waters. Instead the main phototrophs are cyanobacteria, phytoflagellates and chlorophytes. A handful of genera appear to be particularly successful in these non-marine aquatic habitats. Members of the cyanobacterial class Oscillatoriaceae, in particular *Phormidium, Oscillatoria* and *Lyngbya* rorm thick films and mats on the lake sediments and streambeds. Flagellates such as *Crypto-monas* and *Chroomonas* (Cryptophyceae), *Ochromonas* (Chrysophyceae) *Chlamydomonas* (Chlorophyceae) and *Pyramimonas* (Prasinophyceae, see Fig. 3.8) form monospecific population maxima in highly stable lakes. Chlorococcaleans (Chlorophyceae) such as *Ankistrodesmus* and *Schroederia* dominate the plankton of certain maritime lakes.

These phototrophic populations support a wide range of microbial heterotrophs. Data on protozoan and fungal communities are still sparse for maritime as well as continental waters, but like the other microbial groups the diversity of species is much reduced by comparison with temperate regions. For example, the common freshwater fungal group Saprolegniaceae is represented by only one species in Signy Island lakes.

### 7.3.1 Phototrophs

Cyanobacteria occur in all types of freshwater habitat in Antarctica, and are often the microbial biomass dominants in streams, lake sediments and plankton communities. Two genera in Lake Vanda, *Phormidium* and *Synechocystis* dominate the plankton in the deeper strata of the water column, and give rise to a deep chlorophyll *a* maximum in this lake (see Fig. 7.9). *Phormidium* is also planktonic in Lake Miers, 100 km to the south of Lake Vanda, but elsewhere in Antarctica this genus seems a much less important element of the phytoplankton. By contrast, *Phormidium* or closely related members of the Oscillatoriaceae typically dominate the lake benthos and stream epilithon at many sites throughout the region.

Cyanobacterial mats overlying the sediments are an important ecosystem feature of many antarctic lakes and pools. In Dry Valley lakes *Phormidium frigidum* is the dominant in all of the mats, sometimes in association with *Lyngbya martensiana*. The mats also contain diatoms, bacteria, yeasts, protozoa and three metazoan groups: rotifers, tardigrades and nematodes. Five types of mat community (Fig. 7.6) can be recognised in these lakes (Parker & Wharton, 1985):

  (i) *Moat mats.* These occur around the edge of the lake where the ice melts each summer to form a 2–10 m wide moat of open water between the shore and the permanent ice. The communities experience ice-scouring and freeze–thaw cycles, but each summer form thick spongy mats.

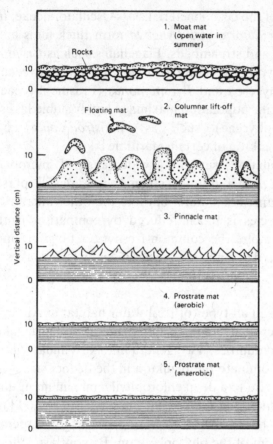

Fig. 7.6. The five main types of benthic cyanobacterial mats in the Dry Valley lakes. Modified from Wharton *et al.* (1983).

(ii) *Columnar lift-off mats.* In the shallow but perennially ice-covered zones of several Dry Valley lakes the mats trap bubbles of oxygen and nitrogen up to 1 cm in diameter. These cause the matrix to grow as an upright, columnar structure. Some of these mats may detach from the sediments and float up to the underside of the ice surface where they become incorporated and eventually move up through the ice cap (see Chapter 2). In Lake Fryxell these columnar mats precipitate calcium carbonate and tend to remain in place with a hard calcite interior.

(iii) *Pinnacle mats.* These grow beneath the thinnest perennial ice of Lake Vanda to depths of at least 30 m. They form cone-shaped structures generally 2–5 cm high, 3–5 cm wide and spaced 6–12 cm apart, containing sand grains and calcite crystals. They resemble the Precambrian fossil stromatolite *Conophyton*.

(iv) *Aerobic prostrate mats*. These flat laminated stromatolites occur at greater depths and also precipitate calcite. This type of mat is rare in certain of the Dry Valley lakes possibly because of the high salt concentrations at depth.

(v) *Anaerobic prostrate mats*. These resemble the type (iv) mats but grow in the deep anoxic basins of some lakes.

Cyanobacterial mats are also a feature of antarctic streams. In the McMurdo Sound region the epilithic communities form black mucilaginous mats up to 10 mm thick, mostly of *Nostoc* (see Fig. 10.7) with a smaller associated biomass of Oscillatoriaceae, and pink, grey-green and orange mats and films composed mostly of Oscillatoriaceae (*Phormidium* spp., *Oscillatoria* spp., *Microcoleus vaginatus*). Some of the rocks have black or dark red surface crusts of *Gloeocapsa* spp. and *Calothrix* sp. (Fig. 7.7). Eukaryotic algae also occur in these streams but generally in much lower biomass levels than the cyanobacterial mats. Long green streamers of *Binuclearia tectorum* develop over the first weeks of flow in certain Dry Valley streams, while *Klebsormidium* coats the streambeds of slightly-brackish flowing waters on Ross Island. The large green alga *Prasiola crispa* is more characteristic of water-flushed soils than streambeds, but another species *P. calophylla* forms green felts in dim-light environments such as beneath the lips of stream boulders. The distribution of these various phototrophic communities can vary with changing nutrient and flow conditions downstream, as well as across the width of the stream (Broady, 1982a; Howard-Williams, *et al.*, 1986a).

A contrasting set of streambed communities has been described from Signy Island streams (Hawes, 1988). Although cyanobacteria (including *Nostoc*, *Phormidium* and *Gloeocapsa*) were observed in many locations the most conspicuous elements were filamentous chlorophytes of the genera *Klebsormidium*, *Zygnema* and *Mougeotia*. These formed luxuriant mats in some of the streams, with filamentous strands more than 30 cm long.

In the streams of both Signy Island and the McMurdo Sound region the standing crop of epilithic biomass can reach very high levels. The Dry Valley mats, however, have very low photosynthetic rates per unit chlorophyll *a* or per unit carbon implying that much of the biologically-derived material is metabolically inactive (see Table 7.1).

The seasonal characteristics of these cyanobacterial mat and film communities contrast markedly with the filamentous chlorophytes in Signy

(a)

(b)

Fig. 7.7. Scanning electron micrographs of benthic cyanobacteria from streams in the Dry Valleys. (*a*) a black epilithic crust from the streambed shown in Fig. 10.7(*b*). The dominant organism is *Calothrix* which has trichomes approximately 8 $\mu$m wide tapering to approximately 2 $\mu$m at their apices. (*b*) an orange mat of Oscillatoriaceae from Fryxell Stream. The trichomes are approximately 6 $\mu$m in diameter.

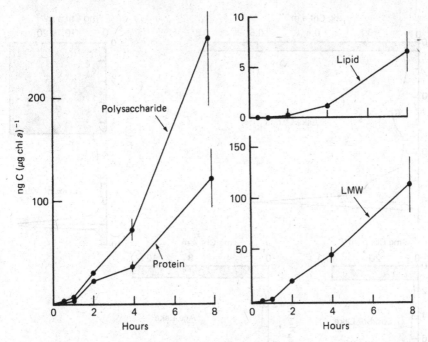

Fig. 7.8. The flow of newly fixed carbon into the end-products of photosynthesis during the first 8 h of rehydration of dry, frozen oscillatoriacean mat from Fryxell Stream, Dry Valleys. Redrawn from Vincent & Howard-Williams (1987).

Island streams. The Dry Valley streambeds retain a high overwintering biomass of cyanobacteria that remains deep-frozen and dry until the next season of streamflow. These overwintering cells begin photosynthetic carbon fixation and the biosynthesis of macromolecules within 60 min of rewetting (Fig. 7.8) but by the end of the season biomass is generally less than double the preflow inoculum (Vincent & Howard-Williams, 1986). By contrast, the Signy Island chlorophytes are mostly killed by the winter freeze-up. A very small residual population begins growth the next season, and steadily grows to a biomass maximum after 2–3 months of streamflow (Hawes, 1988).

Distinct layers of microalgae are often observed in the relatively still waters beneath lake ice. These highly pigmented cells form a well defined 'deep chlorophyll maximum' (DCM) at the depth where nutrient conditions are optimal for growth. In Dry Valley lakes the DCM lies just above the nutrient-rich anoxic zone, which is usually far below the permanent ice cap, for example at 60 m in Lake Vanda (Fig. 7.9). Similar peaks have been recorded in the oxycline of Lake Fryxell, Lake Hoare and Lake Miers

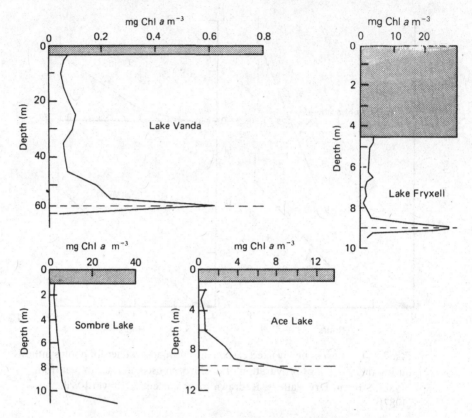

Fig. 7.9. Deep chlorophyll *a* maxima in four antarctic lakes. Redrawn from Vincent & Vincent (1982), Lake Vanda; Vincent (1981), Lake Fryxell; Burch (1987), Ace Lake; and Hawes (1985), Sombre Lake. The dashed line marks the transition from oxic to anoxic water (oxycline).

(southern Victoria Land), Ace Lake and Organic Lake (Vestfold Hills), and at or above the oxycline or near the sediments in clear Signy Island lakes. These populations may slow the transfer of nutrients up into the euphotic zone, and thereby help perpetuate oligotrophy in the overlying water column. DCMs are important in oligotrophic lake and ocean ecosystems throughout the world, and antarctic ice-covered lakes offer model systems in which to define their physiological properties and community dynamics.

Photosynthetic sulphur bacteria are also found in antarctic lakes, especially those derived from seawater where sulphate reduction has generated high concentrations of $H_2S$. A plate of sulphur bacteria typically occurs each year beneath the DCM in Lake Fryxell (Rawley, 1982), Ace Lake (Hand & Burton, 1981), Burton Lake (Burke & Burton, 1988), and probably many other lakes of the region.

### 7.3.2 *Heterotrophic bacteria*

A wide range of bacterial heterotrophs have been identified in antarctic lakes and there is no evidence to date that the polar environment substantially limits the distribution and activity of this microbial group. Direct epifluorescence counts of bacterial cells reveal similar concentrations in antarctic lakes to those observed in temperate latitude waters ($10^5$–$10^6$ cells ml$^{-1}$, e.g. Ellis-Evans 1981a, b; Vincent, 1987). In some of the Dry Valley lakes peak concentrations of bacterioplankton, and also the largest cells, have been observed in the bottommost part of the anoxic zone (Vincent, 1987). In the overlying aerobic zone the highest bacterial cell counts are often observed in the stratum containing the DCM or primary production maximum (Fig. 7.10). As elsewhere, the surficial sediments of antarctic lakes contain orders of magnitude higher concentrations of bacteria than the overlying water column (Ellis–Evans, 1981a, b).

Two broad categories of bacteria occur in the lakes of Signy Island (Ellis-Evans & Wynn-Williams, 1985). Large populations develop in the catchment during the spring thaw and probably exploit the pulses of dissolved organic carbon released by freeze-damaged moss cells (see

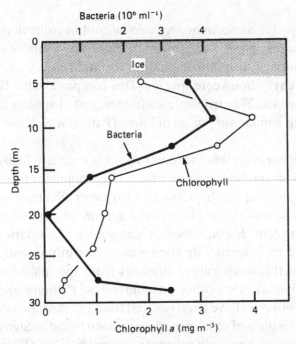

Fig. 7.10. Bimodal distribution of bacterial counts in Lake Hoare (see Fig. 7.1) with a major peak in the stratum containing the DCM. Redrawn from Mikell *et al.* (1984).

Chapter 8). These terrestrial species are washed into the lakes by surface runoff, and mix with the indigenous lake bacteria. Isolates from the lakes included members of the genera *Pseudomonas*, *Vibrio*, *Flavobacterium* and the species *Chromobacterium fluviatile* which occurred in substantial concentrations in Heywood Lake sediments throughout the year. *C. fluviatile* was not found in the water column or in the lake catchment but a related species, *C. lividum* grew to substantial concentrations in the catchment during late summer to autumn. It also appeared in the lake at this time, perhaps completely derived by runoff from the surrounding catchment. The catchment and indigenous lake bacteria differed in their ability to transport organic compounds and in their response to temperature. The terrestrial species were attuned to high concentrations of dissolved organic carbon (DOC) which characterised the moss beds, while the indigenous lakewater species were capable of utilising lower substrate levels. The lake bacterial isolates all had temperature optima for growth below 20 °C, with growth up to 21–24 °C. The catchment isolates had both higher temperature optima (typically >18 °C) and temperature maxima (23–33 °C).

### 7.3.3   *Fungi*

Aquatic phycomycetes are commonly encountered in collections of fungi from antarctic lakes, pools and streams but species diversity is limited. For example, a survey of aquatic habitats in the McMurdo Sound region yielded only four chytridiomycete species. This compared with 17 species from European forests, 27 in tropical rain forests, and 11 species in African salt deserts using similar survey techniques (Paterson & Knox, 1972).

The Oomycete *Aphanomyces* sp. was the most prevalent fungus in Signy Island lakes, particularly in outflow regions where large populations of filamentous green algae develop each summer. Two other filamentous fungi were observed in these waters: the Oomycete *Lagenidium giganteum* and the Hyphochydriomycete *Hyphochytrium catenoides*. This latter species has been reported from temperate zone waters very infrequently yet has been detected in three separate collections from the maritime Antarctic (Ellis-Evans, 1985a). Two other phycomycetes, *Pythium* and *Mucor*, as well as *Aphanomycetes* have been isolated from the shoreline of lakes in the Dry Valleys. Neither of these genera have been found in Signy Island lakes, but they sometimes occur in nearby terrestrial habitats (Pugh & Allsopp, 1982).

A small number of Chytridiomycetes appear to be important microbial elements of coastal antarctic lakes. Sediment samples from Signy Island lakes contained the chytrids *Catenophlyctis variabilis, Rhizophydium* sp., *Rhizophlyctis* sp. and *Phlyctochytrium acuminatum*. Representatives of the genera *Rhisophydium* and *Phlyctochytrium* as well as *Catenophlyctis variabilis* have also been observed from coastal ponds in the McMurdo Sound region. These few species may be present in reasonable abundance. Ellis-Evans' (1985a) study on Signy Island yielded Chytridiomycetes in 67 out of 84 samples. All of the isolates were able to grow at +2 °C and withstood two freeze–thaw cycles. They were unable to grow above 34 °C but the optimum temperatures for growth were in the range 15–20 °C, well above ambient lakewater temperatures.

A number of chytrids have also been observed parasitising algae and other fungi in antarctic lakes. Two species, *Chytriomyces* sp. and *C. willoughbyi*, have been observed as parasites on *Aphanomyces* in Signy Island lake material. A chytrid resembling *Chytridium versatile* parasitised the phytoplankton species *Schroederia* sp. and benthic diatoms in these lakes (Ellis-Evans, 1985a). Infections of this fungus have also been reported on pennate diatoms from ponds in the McMurdo Sound region (Paterson & Knox, 1972).

Many of the species isolated from antarctic lakes represent populations washed in from the surrounding catchment, but like the terrestrial bacteria flushed in with them (see section 7.3.2), they are probably not very active once in the lake. The best examples of this wash-in effect are the yeasts in Signy Island lakes (Ellis-Evans, 1985a). The most common species were basidiomycetes in the genera *Cryptococcus, Leucosporidium, Rhodotorula* and *Candida*. All had temperature optima below 25 °C and grew well at lake temperature as long as the growth medium contained high levels of organic substrates. Identical species are known from terrestrial habitats on Signy Island. The lake yeasts were only present in abundance during spring to early summer when populations were high in the catchment soils. They were not able to compete with bacteria at low levels of DOC and were always present in much lower concentrations than the bacterial populations in the lakes. Even during their spring peak yeast concentrations were several orders of magnitude below the bacterioplankton counts. By contrast in the terrestrial environment, yeast populations rapidly increased during the spring melt and eventually outnumbered bacterial counts for the catchment (see Chapter 8).

### 7.3.4 *Protozoa*

Although protozoan diversity is also reduced in Antarctica a surprisingly large number of species have been reported from continental freshwaters as well as the maritime zone. Artificial substrates consisting of polyurethane foam squares which were incubated in Lake Fryxell, Dry Valleys, for up to five weeks yielded 35 species of protozoa from the water column and 55 species from the littoral zone (Cathey *et al.*, 1981). This diversity seems unusual by comparison with the simple autotrophic community structure in this lake: the phytoplankton contains only four important species while the benthos consists mostly of *Phormidium frigidum* and *Lyngbya martensiana*. However the number of protozoa is still small by comparison with lower latitudes. Investigators using similar colonisation techniques have typically recorded more than 150 protozoan species from temperate lakes and ponds (Cathey *et al.*, 1981).

A detailed series of enrichment assays have been employed at Signy Island to examine the protozoan communities of the freshwater benthos (Hawthorn & Ellis-Evans, 1984). A much larger number of species were recorded from the lakes (75 species) than from the pools (35 species). This probably reflects the much greater environmental stability and persistence of the lake ecosystems, as well as the wider range of microbial habitats. Although the pools differed considerably in their physical and chemical properties they all had a very similar microfauna. As with the microalgae the successful protozoa in this environment are likely to be opportunistic species with wide tolerances to changing physico-chemical conditions. Certain species are excluded, however, from the most extreme environments. The ciliate *Cinetochilum margaritaceum*, for example, is sensitive to free-ammonia and is predictably absent from seal wallows where total ammonium-N concentrations are >1.5 g m$^{-3}$.

Most of the benthic species on Signy Island were small (<150 $\mu$m) and probably fed on bacteria. Ubiquitous species included flagellates such as *Heterochromonas* (a phototroph), *Bodo saltans* and *Heteromita globosa* and ciliates such as *Cinetochilum margaritaceum* and *Aspidisca costata*. Amoebae were well presented in the lake environments (18 species) but the pools rarely contained amoebae and yielded only three species. Flagellate cysts were generally less than 5% of the total flagellate population, whereas encysted forms of rhizopoda accounted for up to 85% of the rhizopod numbers.

Five protozoan genera were enumerated on three occasions in the lake sediments and provide some evidence of a seasonal succession of species. In oligotrophic Sombre Lake dominance shifted from *Bodo* spp. in September when the lake was ice-covered to the phytoflagellate *Heterochromonas* sp. during open water conditions in February. In eutrophic Heywood Lake *Monosigna ovata* accounted for 45% of the November flagellate population, but *Cercomonas longicauda* was the clear dominant

(30% of the total flagellate count) when the lake was free of ice in February. Certain species were restricted to one lake or the other, e.g. *Bodo celer* in Sombre Lake, *Monosigna ovata* in Heywood Lake. The most common ciliate, *Cinetochilum margaritaceum*, occurred in both lakes with no apparent seasonal trends.

These benthic species in Signy Island lakes were restricted to the top few millimetres of sediment and only grew under aerobic conditions. An anaerobic protozoan, however, has been reported from the Vestfold Hills. This enigmatic species, coded 'Eric' by Australian investigators, is unpigmented, lives in the upper anoxic zone of Burton Lake, appears to have certain ultrastructural affinities with the Euglenophyceae and may represent a new genus (H. Burton, pers. comm.). This micro-organism (the anaerobic flagellate in Fig. 7.16) is an important link in the food chain of Burton lake.

Water column protozoa may be important grazers on planktonic microalgae and bacteria. Large ciliates have been recorded in the water column of eutrophic Signy Island lakes (Hawthorn & Ellis-Evans, 1984) and in the DCM of Lake Fryxell of the Dry Valleys. It is known that some phytoflagellates can supplement their nutrition by taking up particulate organic materials, and thus behave like heterotrophic protozoa (e.g. the chrysophyte *Dinobryon*; Bird & Kalff, 1986). This alternate mode of nutrition could be especially important for the phytoplankton during winter darkness (see Chapter 10).

## 7.4 Microbial processes

### 7.4.1 *Photosynthesis*

Like all antarctic ecosystems light is available for microbial photosynthesis for only a portion of the year (see Appendix 2). In shallow ephemeral pools the production season may be especially brief before the pond refreezes or dries up. The seasonal availability of light is further reduced by persistent ice cover. This combination of features results in low annual production rates for lakes throughout Antarctica despite an apparently high phototrophic biomass in some waters. For example, taking the limited production data for ice-covered Lake Fryxell (Vincent, 1981) at 77°S and extrapolating to a 180 d growing season gives a rough upper production estimate of only 7 g C m$^{-2}$ y$^{-1}$. This contrasts with productivities 5–10 times higher for temperate lakes with similar (or lower) chlorophyll *a* concentrations (e.g. 1.4–2.2 mg chl*a* m$^{-3}$ in Lake Huron, with an annual production of *c.* 100 g C m$^{-2}$ (Vollenweider, Munawar & Stadelmann, 1974)). Skua Lake on Ross Island (77°C) contained high levels of planktonic chlorophyll *a* in summer, from 5 to 40 mg m$^{-3}$, with an annual productivity of 45–133 g C m$^{-2}$ y$^{-1}$ (Goldman *et al.*, cited in Priddle *et al.*, 1986). This compares with a similar chlorophyll *a* range in Lake Washington (48 °N) in the mid-1960s, but annual productivities at this time were around 250 g C m$^{-2}$ (Edmondson, 1972). Similarly low annual production values have been reported from lakes in the high Arctic, e.g. 4.1 g C m$^{-2}$ y$^{-1}$ for Char Lake and 11.3 g C m$^{-2}$ y$^{-1}$ for Meretta Lake, both at 74 °N (Kalff & Welch, 1974). At lower latitudes in the Antarctic where the seasonal fluctuations in radiation are reduced (see Appendix 2) annual photosynthetic rates are correspondingly higher. For example, Heywood Lake on Signy Island (60 °S) contains chlorophyll *a* concentrations from 5 to 25 mg m$^{-3}$ with annual production estimates in the range 139–270 g C m$^{-2}$ (Priddle *et al.*, 1986). These values are comparable with temperate Lake Erie (Western Basin): 3–19 mg chl*a* m$^{-3}$ and 210 g C m$^{-2}$ y$^{-1}$ (Vollenweider *et al.*, 1974).

Variations in the thickness of snow and ice cover over antarctic lakes can induce abrupt changes in photosynthesis by the planktonic and benthic phototrophs beneath. In Sombre Lake light availability suddenly rose by more than twenty-fold over September–October because of the loss of overlying snow, coupled with the seasonal rise in radiation. Photosynthetic rates showed a strong positive reponse to this more favourable energy supply, with an increase in both chlorophyll *a* and photosynthetic rates per unit chlorophyll *a* (assimilation number). These variables increased to their highest values when the lake ice completely melted out in late summer and light availability was maximal (Fig. 7.11).

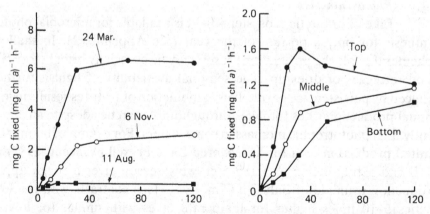

Fig. 7.11. Photosynthesis versus light curves for the phytoplankton in Sombre Lake, Signy Island. The curves on the left are for three different times of year; the curves on the right are for three depths down the water column under the spring ice cover. Redrawn from Hawes (1985).

In various antarctic waters the maximum assimilation numbers for microbial photosynthesis seem unusually low: often $<1$ $\mu$g C $(\mu$g chl$a)^{-1}$ h$^{-1}$, which contrasts with typical values for temperate zone phytoplankton of $3$–$6$ $\mu$g C$(\mu$g chl$a)^{-1}$ h$^{-1}$ (e.g. Parsons & Takahashi, 1973). Similarly low assimilation numbers have been observed in arctic lakes (Table 7.1), and attributed to the low (near 0 °C) ambient temperatures (Rigler, 1978). Temperature does not seem to be the primary variable directly controlling $P_{max}$ per unit chlorophyll in antarctic lakes, for there are large variations in the assimilation number at the same near-freezing temperatures (e.g. Lake Fryxell and Skua Lake, Table 7.1).

Table 7.1. *Assimilation numbers for antarctic inland water communities*

$P_{max}$ = photosynthesis measured in the laboratory or field under saturating light; $A_{max}$ = maximum photosynthesis measured under ambient field conditions ($P_{max}$ not necessarily attained).

| | Ice cover (m) | Temperature (°C) | $P_{max}$ or $A_{max}$ ($\mu$g C ($\mu$g chlorophyll $a)^{-1}$ h$^{-1}$) | Reference |
|---|---|---|---|---|
| **Phytoplankton** | | | | |
| *Signy Island* ($P_{max}$) | | | | |
| Sombre Lake (Aug) | 1 | 1.5 | 0.3 | Hawes (1985) |
| Sombre Lake (Mar) | 0 | 1.5 | 6.7 | Hawes (1985) |
| Heywood Lake (Aug) | 1 | 1.5 | 1.1 | Hawes (1985) |
| Heywood Lake (Mar) | 0 | 1.5 | 11.8 | Hawes (1985) |
| *Ross Island* | | | | |
| Algal Lake | 0 | $<5$ | 10.0 | Goldman *et al.* (1972) |
| Skua Lake | 0 | $<5$ | 7.5 | Goldman *et al.* (1972) |
| *Dry Valley Lakes* ($A_{max}$) | | | | |
| Lake Vanda (3.5 m)[a] | 3.25 | 4.5 | 0.11 | Vincent & Vincent (1982a) |
| Lake Vanda (57.5 m) | 3.25 | 19.0 | 0.14 | Vincent & Vincent (1982a) |
| Lake Fryxell (9.0 m) | 4.5 | 2.0 | 0.11 | Vincent (1981) |
| Lake Joyce (15 m) | 5.0 | 2.0 | 0.01 | Parker *et al.* (1982b) |
| Lake Bonney (12 m) | 4.0 | 0.5 | 0.02 | Parker *et al.* (1982b) |
| **Mat communities** ($P_{max}$) | | | | |
| *Fryxell Stream* | | | | |
| Nostoc | 0 | 0 | 0.17 | Vincent & Howard-Williams (1988) |
| Phormidium | 0 | 0 | 0.94 | Vincent & Howard-Williams (1988) |
| Binuclearia | 0 | 0 | 0.79 | Vincent & Howard-Williams (1988) |

[a] Water depth of collection and incubation.

Conversely, in Dry Valley lakes such as Lake Vanda the assimilation numbers remain extremely low despite relatively warm temperatures in the solar-heated bottom waters of the lakes (Table 7.1).

Two other factors are likely to be responsible for the low measured values. Firstly, lake phytoplankton are likely to become shade-adapted as a result of the persistent snow and ice cover. Such cells increase their light-harvesting capability, which is typically manifested as more cellular pigment per unit photosynthesis, i.e. a lower assimilation number. This effect is strikingly illustrated by the order of magnitude rise in the value of this parameter in Signy Island lakes between the dim-light, sub-ice conditions in winter and the bright-light open water regime in late summer (Fig. 7.11; Table 7.1).

A second factor contributing towards the low assimilation number may be the persistence of photosynthetically inactive 'apparent chlorophyll *a*' associated with senescing cells or chlorophyll *a* decomposition products. Chlorophyllide *a* and phaeophorbide *a*, for example, have very similar properties to chlorophyll *a* and phaeophytin *a* and are not corrected for in standard pigment analyses. The persistence of undecomposed 'apparent chlorophyll *a*' in the cyanobacterial mat communities of Dry Valley streams has been identified as the primary explanation of the low measured assimilation numbers (Vincent & Howard-Williams, 1988). In these communities both the amount of measured chlorophyll *a* and standing stock carbon are extremely high relative to the amount of carbon fixed ($>1$ $\mu$g chl*a* ($\mu$g C fixed h$^{-1}$)$^{-1}$, $>500$ $\mu$g biomass-C ($\mu$g C fixed h$^{-1}$)$^{-1}$), suggesting the persistence of large quantities of biologically-derived material that is not associated with metabolically active cells. A series of temperature-controlled bacterial assays indicated that under warmer conditions much of this material would completely decompose (Vincent & Howard-Williams, 1988). Although these thick, perennial mats of cyanobacteria represent an extreme example, it is possible that undecomposed inactive cells or detritus have similarly produced overestimates of chlorophyll *a* (hence underestimates of assimilation number) in other antarctic cold-water environments.

The physically stable water columns of ice-covered antarctic lakes induce a high level of vertical structure in the species composition, biomass distribution and also photosynthetic performance of the algal community. In the Dry Valley lakes (Vincent, 1987) and certain Vestfold Hills lakes (e.g. Ace Lake, Hand & Burton, 1981) peak photosynthesis for the water column occurs in the DCM region of maximum biomass. The photosynthesis versus light curves can be very different for individual phototrophic communities at different depths, indicating adaptation by the cells to specific light, temperature and nutrient regimes at the various positions down the water column (Fig. 7.11).

### 7.4.2 Bacterial heterotrophy

Like the phototrophs, the microbial heterotrophs of ice-covered lakes may similarly show large variations in population structure and activity with depth. For some of the bacterial populations there is a close temporal and spatial coupling with the phototrophic community. In Lake Vanda, maximum rates of thymidine incorporation into DNA were recorded in the region of the DCM (Vincent *et al.*, 1981). In the Signy Island lakes maximum bacterial counts as well as fastest organic carbon turnover rates followed the seasonal variation in chlorophyll *a* at a slight lag (Ellis-Evans, 1981a, b). For these maritime waters there was evidence of a close reciprocal interaction between the microbial phototrophs and heterotrophs. Photosynthetic rates of Heywood Lake and Sombre Lake water cultures of phytoplankton were significantly stimulated by the addition of lake bacteria, and these effects were eliminated by parallel additions of bacterial antibiotics (Ellis-Evans, 1985c).

High dissolved oxygen tensions may be a factor influencing heterotrophic processes in ice-covered lakes. In Lake Hoare (Dry Valleys), oxygen concentrations are often more than 40 g m$^{-3}$, but glucose assimilation and catabolism by the planktonic bacteria seemed unaffected by these elevated concentrations. By contrast the same processes in benthic microbial mats from Lake Hoare, and also by bacterioplankton from a temperate latitude lake, were found to be significantly inhibited by similarly high concentrations of dissolved oxygen (Mikell, Parker & Simmons, 1984). It is also possible that these unusual oxygen conditions have other effects on microbial activity. The Lake Hoare bacterioplankton readily catabolise glycolate, and this substrate could potentially be released during microalgal photorespiration, a process favoured by high oxygen tensions (Parker & Simmons, cited in Mikell *et al.*, 1984).

### 7.4.3 Nutrient cycling

The ice-capped lakes of Antarctica are ideal natural systems for experimental studies of aquatic nutrient cycling. The physically stable water columns in these ecosystems produce strong vertical gradients in the distribution of environmental properties which in turn favour vertical segregation of microbial populations. This vertical community structure has had a reciprocal influence on the chemical environment; in many of the lakes sharp gradients in the concentration of specific compounds have been created by the localised activity of specific micro-organisms.

Lake Vanda in the Dry Valleys of southern Victoria Land (see Appendix 1) has attracted special interest in microbial nutrient cycling investigations because of its deep stratified water column, perennial bottom-water anoxia, permanent ice cover (*c.* 3.5 m thick) and relative accessibility to investigators from many nations. The lake has a number of highly unusual features (e.g. increasing temperatures with depth, to 24 °C; see Vincent, 1987). Microbial studies have been entirely limited to the summer logistics season. Nevertheless the nutrient transfers in Lake Vanda are illustrative of microbial cycling processes that operate in water bodies throughout Antarctica, and this lake has therefore been selected as the focus for the following descriptions.

## Phosphorus

Phosphorus (P) availability exerts an overall control on phototrophic biomass in Lake Vanda (Vincent & Vincent, 1982a) and the processes which control the abundance of this element are therefore pre-eminently important. Phosphorus limitation has also been identified elsewhere in Antarctica (e.g. Sombre Lake, Signy Island; Hawes, 1985), however there can be large differences in the relative importance of nitrogen and phosphorus even in nearby lakes (e.g. N-deficiency in Heywood Lake, Signy Island; Hawes, 1985).

Dissolved *reactive phosphorus* (phosphate and other low molecular weight phosphorus species) is in very low concentration throughout the aerobic water column, typically $\leq 0.5$ mg P m$^{-3}$. Phosphorus levels rise abruptly across the oxycline, and this increased availability of the limiting element probably accounts for the population maximum of cyanobacteria which comprise the DCM (Fig. 7.9). Several biogeochemical processes control the transfer of phosphorus across the oxycline. Phosphate is produced by the bacterial mineralisation of microbial (especially phytoplankton) cells that have settled from the overlying waters and may be further released from adsorption sites on hydrated oxides that are reduced in the anoxic zone. Much of this phosphate may be lost to the sediments by the formation of calcium hydroxyapatite ($Ca_5(PO_4)_3OH$): the ion activity product for this mineral in the bottom waters of Lake Vanda is several orders of magnitude higher than the equilibrium constant indicating that the anoxic zone may act as a net chemical sink for phosphorus (Canfield and Green, 1985).

Some of the phosphate diffuses up from the anoxic zone, but turbulence is believed to be absent from these bottom-most layers because of the lack of direct, wind-induced mixing (permanent ice-cover) and the strong salinity gradients (*c.* 8‰ at 50 m to *c.* 80‰ at 68 m). The flux ($F_p$) is therefore controlled by slow, molecular diffusion (Canfield & Green, 1985):

$$F_p = D_p \frac{\partial C_p}{\partial z} \qquad (7.9)$$

where $D_p$ = molecular diffusivity for phosphate
$\qquad\quad = 7.15 \times 10^{-6} \text{ cm}^2 \text{ s}^{-1}$

$\frac{\partial C_p}{\partial z}$ = concentration gradient for phosphate across the oxycline.

From measured values of the phosphate concentration gradient in Lake Vanda and extrapolating the very limited photosynthetic production data to a hypothetical six month growing season, it is apparent that the DCM in Lake Vanda has the potential to completely consume this upward flux of phosphate (about 14.5 mol $y^{-1}$) and thereby severely restrict the availability of phosphorus to the phototrophs living under more favourable irradiance conditions higher in the water column (Canfield & Green, 1985).

The upper communities of phototrophs probably gain most of their phosphorus from the single inflow to Lake Vanda, the Onyx River. Canfield & Green (1985) estimate that about 80 mol of dissolved inorganic phosphorus enter the lake from this source each year, at least a factor of five greater than the amount delivered by molecular diffusion from the anoxic zone. Even this input, however, is low relative to the potential phototrophic demand for phosphorus (about 500 mol P $y^{-1}$ from the carbon fixation data), and phosphorus recycling rates must be rapid at all depths, including within the region of the DCM.

A major unknown in the phosphorus cycle (and all other nutrient cycles) in Lake Vanda is the role of the benthic cyanobacterial mats. It is likely that there is a tight recycling of nutrients within these mixed assemblages of phototrophs and heterotrophs. Given their almost continuous cover in parts of the lake (Wharton, Parker & Simmons, 1983) the benthic mats may effectively restrict the diffusion of nutrients out from the sediments and thereby help contribute towards the oligotrophic status of the water column.

*Nitrogen*

A much more complex series of microbial pathways control the distribution and forms of nitrogen in Lake Vanda. Ammonium is regenerated in the anoxic zone, primarily by sulphate reduction (see below). There are not the same geochemical controls on nitrogen as there are on phosphorus and much of this $NH_4$-nitrogen is free to diffuse up into the euphotic zone. Calculating this flux as for phosphorus Canfield & Green (1985) estimate that 8800 mol N are delivered to the aerobic water column from this source, while only 1200 mol N enter the lake in the Onyx River inflow each year. The DCM is a much less effective trap for this bottom-water $NH_4$ than it is for the diffusive flux of phosphorus (Fig. 7.12) and some of the nitrogen therefore diffuses higher into the water column. A portion of this ammonia is oxidised by nitrifying bacteria to nitrite, nitrous oxide and nitrate, and these end-products have accumulated as a distinct band of oxidised N in the depth region 50–60 m (Fig. 7.13). Bacterial assays have confirmed that nitrification rates are maximal in this stratum. Peak activity occurred several metres above the oxycline (Vincent *et al.*, 1981), perhaps indicating that the nitrifying bacteria were inhibited by increasing $NH_3$, $H_2S$ and/or salinity at greater depths. The oxidised nitrogen compounds

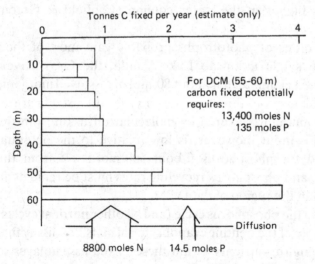

Fig. 7.12. Depth distribution of photosynthetic production in Lake Vanda and the diffusive flux of inorganic N and P from the anoxic zone. The fixation rates have been extrapolated from two 24 h production estimates to a 180 day growing season, and have been adjusted to whole-lake estimates using the depth–volume curve for this lake. The potential DCM demand for N and P has been converted from carbon fixation in the 55 to 60 m stratum using measured stoichiometries. Derived from Canfield & Green (1985) and original data.

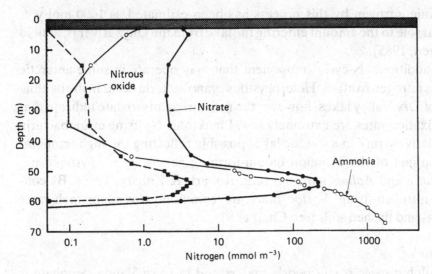

Fig. 7.13. The distribution of nitrous oxide, nitrate and ammonium in Lake Vanda. Derived from Vincent *et al.* (1981).

diffuse away from the band of nitrification and may be very slowly assimilated (at slower rates than ammonium) by the phytoplankton. Some of this oxidised nitrogen diffuses down into the anoxic zone where it is consumed by denitrifying bacteria and converted to $N_2$. The population of denitrifiers is restricted to a very narrow (*c.* 2.5 m) band immediately beneath the DCM (Vincent *et al.*, 1981). The net loss of combined

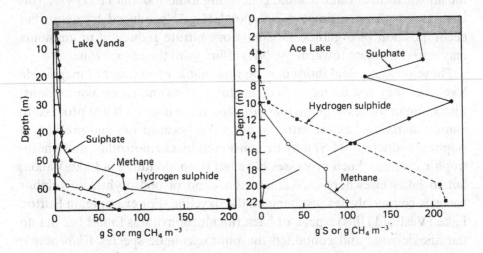

Fig. 7.14. The distribution of sulphate, $H_2S$ and methane in Lake Vanda (from Torii (1975), and original data) and in Ace Lake (from Hand & Burton, 1981).

inorganic nitrogen by this process has been estimated as $1600$ mol $y^{-1}$, comparable to the amount entering the lake from the Onyx River (Canfield & Green, 1985).

An additional N-cycle component that may operate in some antarctic lakes is nitrogen fixation. Heterocystous cyanobacteria occur in the benthic mats of Dry Valley lakes, however the measurements to date indicate that their fixation rates are extremely low. Planktonic $N_2$-fixing cyanobacteria are relatively rare in antarctic lakes possibly reflecting the high temperature optima of the common bloom-forming genera such as *Anabaena, Nodularia* and *Aphanizomenon* (e.g. Robarts & Zohary, 1987). By contrast, nitrogen-fixing *Nostoc* mats are commonly found in meltwater streams and flushed soils (see Chapter 8).

## Sulphur

The sulphur cycle is still poorly understood in Lake Vanda. Sulphate is distributed throughout the water column, rising to high concentrations in the saline bottom waters (Fig. 7.14). A very small fraction of this sulphur is assimilated and reduced by the phytoplankton and associated microbial cells throughout the euphotic zone. In the anoxic bottom waters, sulphate-reducing bacteria reduce $SO_4$ to $H_2S$, and high concentrations of $H_2S$ have accumulated at the lowermost depths (Fig. 7.14). The sulphate-reducing bacteria are heterotrophic and in the process of decomposing organic substrates they release large quantities of ammonia. Canfield & Green (1985) have concluded that these bacteria are the dominant ammonifiers in the anoxic zone of Lake Vanda, generating some $7000$ mol $NH_4$ $y^{-1}$. This compares with an estimated $300$ mol $NH_4$ $y^{-1}$ produced by denitrifier decomposition of organics. Dissimilatory nitrate reduction to ammonia may also contribute towards the $NH_4$ efflux from the anoxic zone.

These various calculations may be reasonable at an order of magnitude level, but they rest on unverified stoichiometries and many assumptions; for example; the assumption that diffusion is the dominant loss process. In many antarctic lakes the latter is not valid because the end-product of sulphate reduction, $H_2S$, may be consumed by chemotrophic and phototrophic bacteria. Such processes have not been identified in Lake Vanda, but in lakes elsewhere in Antarctica a band of photosynthetic sulphur bacteria commonly lies just beneath the oxycline. For example, in Burton Lake (Vestfold Hills) a peak of bacteriochlorophyll was found 1–2 m into the anoxic zone and contained the photosynthetic species *Chlorobium vibrioforme* and *Chlorobium limnocola*, with associated populations of *Rhodopseudomonas palustris* and *Thiocapsa roseopersicina* (Burke &

Burton, 1988). Similar populations have been observed in Ace Lake (Vestfold Hills) and Lake Fryxell (Dry Valleys). At least one of the lakes in the Vestfold Hills region lacks $H_2S$ in its anoxic zone despite high concentrations of sulphate. In Organic Lake the high salinities (*c.* 140 g Cl $l^{-1}$) coupled with low water temperatures ($-5\,°C$) appear to have excluded $SO_4$-reducing bacteria. However, high concentrations of another reduced-sulphur compound, dimethyl sulphide, accumulate in the anoxic zone (up to 98 mg S $l^{-1}$). Franzmann *et al.*(1987b) have isolated a Gram-negative bacterium from Organic Lake that is capable of degrading sulphur-containing compounds to dimethyl sulphide and which is possibly responsible for the unusually high concentrations of this compound in the lake.

## Methanogenesis and sulphate reduction

There can be strong positive or negative interactions between different microbial groups of the same or even different elemental cycles. Methanogenic bacteria have been identified in many types of anoxic, inland water habitat in Antarctica, including Lake Vanda, generally by the end-product of their metabolism, methane. An important control on their activity at all latitudes is sulphate availability. Under conditions of low redox tensions and high $SO_4^{2-}$ concentration sulphate-reducing bacteria outcompete methanogens for acetate and hydrogen (Ward & Winfrey, 1985). These effects are especially apparent in Antarctica where many of the lakes have received large inputs of marine-derived salts, including sulphate.

There are three broad categories of anoxic water that can be distinguished by their $SO_4$ characteristics and corresponding methanogen activity:

(i) *Freshwaters low in sulphate.* The Signy Island lakes for example have moderate $SO_4$ levels and methanogenic activity becomes increasingly important with increasing anoxia, which in turn is related to trophic state. Even in oligotrophic lakes, however, this anaerobic process has been measured in the sediments or in small pockets of anoxic water. In Sombre Lake methanogenesis consumes 13% of the annually sedimented particulate carbon (Ellis-Evans, 1984).

(ii) *Saline waters high in sulphate.* The Dry Valley lakes, such as Lake Vanda, have sulphate-rich anoxic bottom waters in which methane accumulation is relatively slow (Fig. 7.14).

(iii) *Saline waters low in sulphate.* In certain lakes in the Vestfold Hills sulphate-reducing bacteria have completely consumed the $SO_4^{2-}$ in the anoxic zone, and conditions are highly favourable for methane production (e.g. Ace Lake, Fig. 7.14).

### Iron and manganese

Dissolved and particulate iron and manganese have been measured in relatively low concentrations throughout the aerobic water column of Lake Vanda, but they both abruptly increase in the anoxic bottom waters (Fig. 7.15). This distribution is primarily controlled by chemical rather than microbiological processes. A small proportion of the iron and manganese delivered to the upper waters of the lake is removed by microalgae and cyanobacteria which gradually sink into deeper waters. Green *et al.* (1986) calculate that only 10% of the iron and 2.8% of the manganese are potentially lost from the upper lake (4–48 m) in this way. The dominant loss process is the settling of oxides which have formed in the oxygenated upper waters. These sink into the anoxic zone where the decreased pH (5.6 at 65 m, cf. 8.0 in the upper lake) and pE favours dissolution:

$$Fe(OH)_3 + 3H^+ + e^- = Fe^{2+} + 3H_2O$$
$$MnO_2 + 2H^+ = \tfrac{1}{2}O_2 + Mn^{2+} + H_2O$$

The dissolved $Fe^{2+}$ and $Mn^{2+}$ are then free to diffuse upward across the oxycline where they are reprecipitated as oxides that settle back into the anoxic zone. In the lowermost depths of the anoxic zone $Fe^{2+}$ is stripped from solution as iron sulphide, and probably accumulates in the sediments as pyrite ($FeS_2$).

The change in concentration of a chemical species ($C$) in a lake may be approximated by the relationship (Green *et al.*, 1986):

$$V\frac{dC'}{dt} = \phi_i C_i - \phi_o C - kCV \tag{7.10}$$

where     $V$ = volume of the lake
$\phi_i C_i$ = inflow rate × average inflow concentration
$\phi_o C$ = outflow rate × average lake concentration
$kCV$ = internal loss by biological and chemical processes, where $k$ is a first-order removal rate constant.

For Lake Vanda, which has no outflow and a well defined single inflow, the Onyx River, the average time an element spends in the lake (residence time, $t$, where $t = 1/k$) assuming steady state is given by:

$$t = CV/(\phi_i C_i) \tag{7.11}$$

For the upper water column (4–48 m) of Lake Vanda, and defining $C_i$ as the measured dissolved metal content of the Onyx River, this calculation gives $t$ estimates of 47 y for iron, 55 y for manganese, 82 y for cadmium and 237 y for copper. These residence times are all extremely long (typically <1 y in temperate lakes) and reflect the ultra-oligotrophy of Lake Vanda with extremely low microbial populations to biologically scavenge the dissolved metals (Green *et al.*, 1986). The assumption of steady state in these calculations, however, is a weakness that in view of the recent major rises in lake level will need to be fully addressed in future studies (Chinn & McSaveney, 1987).

Elsewhere in Antarctica the iron and manganese cycles may be more closely coupled to microbiological processes. Iron occurs in very high concentrations in Signy Island lakes, for example up to 18 000 mg dissolved Fe m$^{-3}$ in the bottom waters of Sombre Lake (Gallagher, 1985; cf. Fig. 7.15). The sediment-released iron precipitates out as oxyhydroxides which adsorb phosphorus and return it to the sediments. This process may contribute towards the phosphorus limitation of phototrophic growth and biomass in some of these lakes in an analogous way to the calcium hydroxypatite effects in Lake Vanda (see above).

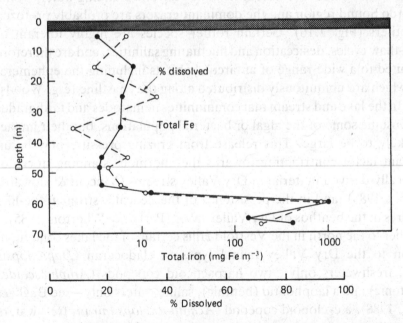

Fig. 7.15. The distribution of total iron (unfiltered samples) in Lake Vanda and the percentage of this Fe that passed through a 0.45 μm filter (% dissolved). Derived from Green *et al.* (1986).

Oxidised Fe may also provide a terminal electron acceptor for iron-reducing bacteria in the anoxic zone of some antarctic lakes. Such bacteria have been cultured from the sediments of Sombre Lake and these isolates preferentially utilise $Fe^{3+}$ oxyhydroxides that would be sinking out from the overlying water column. Thermodynamic considerations indicate that such bacteria should outcompete both methanogens and sulphate-reducers, and could therefore play a major role in the anaerobic decomposition of organic carbon (Ellis-Evans & Lemon, 1988).

### 7.5    Trophic structure

The trophic structure of antarctic inland water ecosystems is more complex than the simple terrestrial food chains of continental Antarctica, but much less so than the food web of the Southern Ocean. Comparative faunal data are only available from three locations: Signy Island (60 °S), Vestfold Hills (68 °S) and the McMurdo Sound region (78 °S). These limited observations imply that trophic complexity in the lakes, streams and pools decreases southward. In certain parts of Antarctica there are unusual transitional lakes or lagoons that retain a tenuous connection with the sea which allows marine species to additionally enter the food web.

Crustacean animals are absent from flowing and standing waters of the McMurdo Sound region and the dominant grazers are probably protozoa and rotifers (Fig. 7.16). Certain rotifer species are highly tolerant of freeze–thaw cycles, desiccation and fluctuating salinities, and are therefore well suited to a wide range of antarctic habitats including the ephemeral pools which are ubiquitously distributed along the coastline (e.g. Woods, 1976). In the lake and stream mat communities nematodes and tardigrades may consume some of the algal or bacterial populations, but their impact is unlikely to be large. This release from grazing pressure may be an important factor contributing towards the enormous standing stocks of biologically-derived material in Dry Valley streams (Vincent & Howard-Williams, 1988), and in the persistence of the delicate, stromatolite-like structures in the benthos of Dry Valley lakes (Parker & Wharton, 1985).

Further to the north in the Vestfold Hills certain of the lakes contain, in addition to the Dry Valley faunal types, a cladoceran (*Daphniopsis studeri*; freshwaters only), two harpacticoid copepods (*Amphiascoides* (planktonic) and a laophontid (benthic); saline waters only – see Bayly & Eslake, 1988), a cyclopoid copepod (*Acanthocyclops mirnii*; freshwaters only) and two calanoid copepod species (*Drepanopus bispinosus* and *Paralabidocera antarctica*; saline lakes and the sea). Many of the Vestfold lakes are saline and support organisms that are more commonly found in the sea. Hypersaline Organic Lake, for example, contains a choanoflagellate closely resembling *Acanthocorbis unguiculata*, a species found in various marine locations including the inshore sea about 16 km away from the lake (van den Hoff & Franzmann, 1986). A relatively complex food web has been described for Burton Lake, which receives irregular inputs of seawater during high tides (Fig. 7.16). Four marine invertebrate species inhabited this lagoon: the calanoid copepod *Drepanopus bispinosus*, which attained maximum densities just above the deep chlorophyll maximum, the calanoid *Paralabidocera antarctica* at low densities, the hydromedusa *Rathkea lizzioides*, and an unidentified ctenophore. The marine ice-fish (*Pagothenia borchgrevinki*) is also known to gain access to this lake, but it is unlikely that this species maintains an overwintering population (Bayly, 1986). The hydromedusa was found to occur in the aerobic zone and also to enter the upper stratum of the anoxic zone. In this lagoon there may be an unusual anaerobic food chain from phototrophs (photosynthetic sulphur bacteria) to primary consumers (the heterotrophic flagellate) to secondary consumers (*Rathkea*).

The lakes and pools of Signy Island contain a wider range of freshwater faunal groups. Two planktonic crustaceans, the calanoid copepod *Pseudoboeckella poppei* (especially early copepodite stages) and the anostracan *Branchinecta gaini* feed on the phytoplankton. In Heywood Lake, for example, the anostracans hatch soon after the spring bloom of chlorophytes. They achieve very high population densities (up to 40 000 animals $m^{-3}$) that may consume 40% of the peak phytoplankton standing stock (Hawes, 1985). A large copepod, *Parabroteas sarsi* preys on the copepodite stages of *Pseudoboeckella poppei*. Greatest faunal diversity is found in the benthos which includes, in addition to the species mentioned, three cladocera, two ostracods and an annelid (Heywood, 1984). These animal populations are supported by a more complex community of benthic phototrophs (and associated detrital microflora) than in the Dry Valley lakes, with luxuriant stands of moss in addition to cyanobacterial and chlorophyte mats and films.

Rotifers are ubiquitously distributed throughout maritime as well as continental waters of Antarctica, but comparative data are still sparse and regional variations in diversity, population structure and the grazing impact of these small animals are still poorly understood. The most detailed studies have been on Signy Island where some 28 species have been identified (Dartnall & Hollowday, 1985). Thirty of these were found in the lakes and eight in the pools (defined as bodies of water which seasonally freeze solid) with no species common to both habitats. The

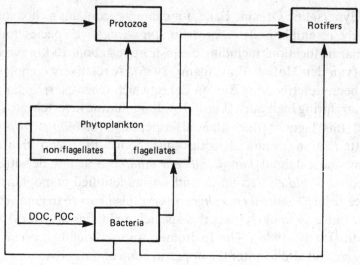

**a**

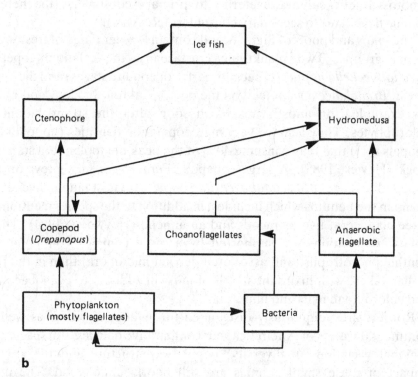

**b**

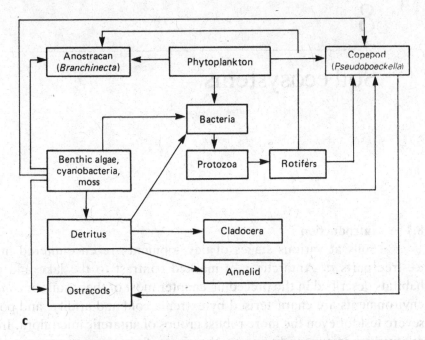

c

Fig. 7.16. Trophic relationships in three antarctic lakes. The Lake Vanda pathways (a) are derived from Cathey *et al.* (1981); (b) Burton Lake as shown for the water column only by Bayly (1986); and, (c) Sombre Lake from Heywood (1984).

greatest species diversity of rotifers was recorded in the moss-containing lakes, further indicating the importance of benthic vegetation in structuring the community at higher trophic levels.

# 8

## Soil ecosystems

### 8.1    Introduction

Soils at various stages of development are encountered in the
ice-free parts of Antarctica. In marked contrast to the lake and pool
habitats described in the preceding chapter most of these soils or *edaphic*
environments are characterised by extreme cold and aridity, and pose a
severe test for even the more robust groups of antarctic microflora. In the
relatively benign conditions at the coast the soils may contain high
concentrations of organic nutrients and a high water content but even here
the summer growing season is marked by rapid freeze–thaw cycles and
concomitantly large shifts in moisture availability. In both inland and
coastal soils the concentrations of soluble salts may rise to levels that
further restrict microbial life.

Despite an apparently hostile regime of low temperature, high salinity
and low moisture content antarctic soils harbour a variety of micro-
organisms. These resilient communities are often dominated by microalgae although fungi, bacteria, actinomycetes and protozoans may be
locally abundant. Yeasts have been isolated from many antarctic soils, and
include species of *Cryptococcus* that are apparently endemic to parts of
the region. Important bacteria include *Micrococcus*, especially in the
organically rich coastal soils, halotolerant *Planococcus* in some saline
continental soils, and the highly-pigmented genus *Chromobacterium* that
occurs in peat deposits at pH as low as 3.7.

Cyanobacteria are usually the dominant phototrophs, but the number of
genera of these and the algal groups are much reduced by comparison with
temperate latitude soils. In parts of the region, especially the maritime
zone (western side of the Antarctic Peninsula, South Orkney Islands,
South Shetland Islands, see Appendix 1) mosses and lichens are the more
important primary producers and these provide a favourable environment
for a wide range of microbial epiphytes. Although a portion of this

phototrophic biomass flows on to microbial heterotrophs and to herbivores such as springtails and mites, both of these groups of consumer organisms are limited by low temperatures and moisture availability, and large standing stocks of undecomposed algae or moss have accumulated at some locations.

## 8.2 The environment

### 8.2.1 *Regional soil types*

Three broad groups of soils can be distinguished in the Antarctic on the basis of climate, organic carbon supply and moisture availability (Claridge and Campbell, 1985). High altitude soils in the Trans-Antarctic Mountains provide the least hospitable environment for microbial growth. Temperatures rise above freezing for only brief periods each year and the permanently frozen ground (*permafrost*) persists close to the surface. The ancient *ahumic* soils of this region date back several million years and are typically covered by a stone pavement (lag gravel) of the more resistant rocks. Below this lies a thin (1–4 cm) horizon of fine gravel and sand derived from the decomposition of the surface rocks. This overlies the 'soil proper' which consists of oxidised and stained fine-textured material with up to 4% clay and accumulations of salts as either a distinct horizon or as small scattered pockets. Chemical weathering is minimal in these soils. The ferromagnesian minerals of basic igneous rocks are the first to decompose. This process releases iron which precipitates on the surface of the more stable soil materials to produce a red-brown stain. Sandstones, where present, provide an important source of clay minerals. These clay micas may hydrate to form vermiculites, or under conditions of high pH and salinity may be converted to smectites. In the Pensacola Mountains (*c.* 83 °S, 50 °W) which lie at one end of the Trans-Antarctic Mountain Range the clay content was typically less than 2% indicating that the dominant weathering processes were physical rather than chemical (Parker *et al.*, 1982a).

In the ice-free regions at the margins of the continent such as the Dry Valleys of southern Victoria Land, Bunger Hills, Vestfold Hills and Enderby Land, the climate is much warmer with summer temperatures to 5 °C or above. The soils experience a slightly greater availability of moisture and deeper penetration of chemical weathering that can be more intense than in the Trans-Antarctic Mountains. Salts are abundant especially on the surface and there may be relatively large quantities of clay. In areas with moss, lichen or algal growth the soil may contain a surface 2–3 cm of humus-containing sandy loam overlying up to 10 cm of sand with traces of organics. These young humic soils are typically grey, although the older soils of Enderby Land have a reddish hue. The higher temperatures and moisture availability in this soil zone relative to the Trans-Antarctic Mountains result in considerable frost-heaving and the soils tend not to have distinct horizons.

A third soil group incorporates those soils which lie at the continental coastline or on offshore islands. They receive materials directly from seaspray or from the birds and mammals which feed in the sea. This category includes soils in the maritime climatic zone, where drizzle is frequent during summer and mean daily temperatures are above 0 °C in January (Appendix 1). In these cold oceanic conditions the soils have a relatively high moisture content and therefore provide more favourable conditions for microbial growth. However, they also regularly experience freeze–thaw cycles in summer which physically disrupt the profiles and prevent the development of well-defined horizons. Chemical weathering proceeds much more rapidly than in the other two soil regions and they may contain up to 10% of clay-sized material. The organic nitrogen and carbon content is high in areas of birdlife or plant growth, and the exchange capacity and concentration of cations, mostly calcium and magnesium may also be high. The soils within penguin rookeries, termed *ornithogenic* soils (Ugolini, 1972), contain especially rich levels of nitrogen, with C:N–ratios often less than two (e.g. Speir & Cowling, 1984).

Mosses, lichens and algae can produce patches up to several hundred metres wide at certain locations in the maritime zone, and large quantities of organic matter may build up beneath the vegetation. This material decomposes very slowly which leads to an accumulation of peat that is permanently frozen and up to several metres thick. Peat deposits up to 2.05 m deep on the South Orkney Islands and 3.4 m deep in the South Shetland Islands have been dated at 2000–5000 years old. Much more recent (500 years) peat deposits up to 1.25 m thick have been recorded on Litchfield Island, close to the Antarctic Peninsula (reviewed in Lewis Smith, 1984). These organically-rich soils of the maritime zone contrast sharply with the Trans-Antarctic Mountain soils, but resemble soils from the north polar deserts.

### 8.2.2   *Moisture content*

Extreme aridity is the single most important feature affecting the suitability of continental antarctic soils as a microbial habitat. The vast majority of Dry Valley soils, for example, have moisture contents comparable with temperate latitude deserts. Victoria Valley soil contained 0.3% $H_2O$, and at many southern Victoria Land sites the moisture content was around 1%; these values compare with the water content of Arizonan and Australian desert sands, 0.2% and 0.7% water, respectively (Atlas *et al.*, 1978). Snowfalls are relatively rare in the Dry Valleys but may be extremely important in stimulating brief periods of microbial activity at irregular intervals. Hollows and ridges that trap the wind-blown snow are presumably more favourable habitats. Contraction-crack polygons, for example, are common throughout periglacial regions including the ice-free parts of Antarctica (Sugden, 1982) and give rise to microtopographic features that cause local variations in snow and meltwater distribution (Fig. 8.1).

Much higher levels of moisture can occur in marine-influenced soils. The water content of the ornithogenic soils at Cape Bird, Ross Island, for

Fig. 8.1. Entrance to the Taylor Valley, southern Victoria Land, looking back towards New Harbour (McMurdo Sound) which is still covered by sea ice at this time of year (November). Apart from isolated patches of mosses and lichens the soils of these so-called Dry Valleys are devoid of vegetation. Microtopographic features including contraction cracks influence the distribution of wind-blown snow and thus the availability of meltwater to the soil microbial communities.

example, ranged from 6% at a site away from the Adélie penguin nesting area, to 43% in a recently abandoned nesting site (Ramsay, 1983). The peat moss habitats in the maritime zone also retain a high moisture content – for example wet weights of up to six times the dry weight in *Drepanocladus* stands in Paradise Harbour (Wynn-Williams, 1984). Again local topography has a major influence on water level: on Signy Island a free-draining, sloping site attained a maximum moisture content of about 1000%, while a site in a hollow showed values in excess of 2600%. These two sites were vegetated by completely different moss assemblages (Walton, 1984).

### 8.2.3  *Temperature*

Temperature exerts an important indirect as well as direct influence on antarctic soil processes. Low temperatures directly inhibit microbial metabolic activities as well as soil weathering reactions. Additionally, the freezing environment restricts the availability of water in liquid form which further limits chemical weathering and biological processes.

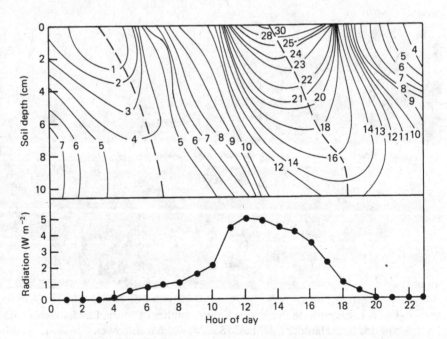

Fig. 8.2. Diurnal heating and cooling of soil during mid-summer on Signy Island. The broken lines mark the times down the profile of temperature minima and maxima. The lower curve shows the 24 hour radiation cycle on the data of measurements at this wet, moss-covered site. Redrawn from Walton (1984).

Antarctic soils may experience considerable diel fluctuations, often passing across their freezing point several times' in a day. In the ornithogenic soils at Cape Bird, surface temperatures varied from <0–10 °C over the 24 h cycle (Ramsay, 1983). Rapid changes in ground heat flux have been recorded over periods of short-term weather variation in the Dry Valleys (Thompson *et al.*, 1971). Strong diurnal temperature variations are also observed in the vegetated parts of the maritime zone. Walton (1982) has shown up to a 30 °C variation over 24 h at the surface of a wet moss stand on Signy Island (Fig. 8.2) with the penetration of maximum temperatures down into the soil profile at 2–3 cm h$^{-1}$.

At the other thermal extreme, steaming hot ground is a feature of the small number of thermal sites throughout the antarctic region. Soil temperatures on the top of Mt Melbourne (about 2700 m ASL) range up to +40 °C despite summer air temperatures of about −20 °C. Bryophytes and microalgae are locally abundant in an area of <200 m$^2$ where the soils are warm and moist. Other geothermal sites in Antarctica are in the South Shetland Islands (Deception Island), the South Sandwich Islands, Ross Island (Mt Erebus), Heard Island (Big Ben), Bouvetøya, and possibly in Marie Byrd Land.

### 8.2.4 *Snow cover*

As in other antarctic ecosystems, snow and·ice cover affect a wide range of environmental variables that have a major influence on the microbiota. Snow provides a reservoir of water that during summer melting may dictate the distribution and abundance of microbial communities. Snow insulates the ground during winter and may allow soil temperatures to persist at values well above the daily air minima, but the high albedo of snow (about 90% for fresh snow, compared with about 10–20% for bare ground, see Appendix 2) retards the rate at which the soil can warm during spring and early summer. On Signy Island, for example, the surface temperature of a moss stand abruptly rose from 1 °C under the snow to 24 °C three days later after the snow had melted (Walton, 1982).

Snow cover, because of its high albedo and attenuation coefficient, also has a major effect on the availability of light for soil phototrophs. On Signy Island, a major increase in photosynthetically available radiation (PAR) at the soil surface was associated with the thawing and refreezing of the snow into a much more transparent combination of snow and ice (Fig. 8.3). The ice layer that formed at the base of the snow pack had an attenuation coefficient for PAR of 0.5 m$^{-1}$, compared with 7.9 m$^{-1}$ for the fresh snow (Walton, 1984).

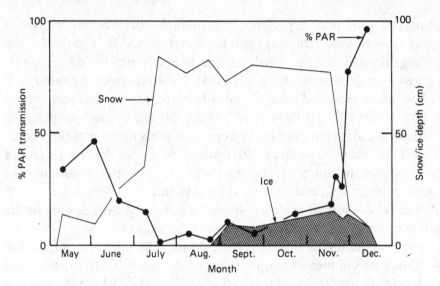

Fig. 8.3. Changes in the penetration of light (photosynthetically available radiation, PAR) to the soil surface with the seasonal change in snow and ice cover, Signy Island. Redrawn from Walton (1984).

### 8.2.5   Salinity

From the pioneer work of Jensen on Shackleton's British Antarctic Expedition of 1907–09 it has long been appreciated that antarctic soils have an unusually high salt content. These salts include calcite (calcium carbonate) and gypsum (calcium sulphate) often as surface encrustations, as well as chlorides, nitrates and other sulphates. They may occur under rocks resting on the soil surface, distributed throughout the profile, or concentrated into discrete horizons either as pure salt or as salt-cemented layers. This accumulation of soluble salts is a feature of hot as well as cold desert regions in which evaporation exceeds precipitation and where the soils are not flushed by stream or meltwater.

The salts are derived from a variety of sources including rock weathering, seaspray, seawater evaporation and atmospheric aerosols. In ornithogenic soils, penguin nasal secretions and ammonium derived from the decomposition of penguin guano, as well as seaspray, are the important sources of salt. Organic salts may also accumulate in the ornithogenic soils: for example fine needle-like spicules of uric acid are commonly detected in the soils and ponds at Cape Bird, Ross Island.

A variety of salt origins were identified in a comprehensive survey of soils along the Trans-Antarctic Mountains (Claridge & Campbell, 1977; Fig. 8.4). Calcium and magnesium were mostly derived from the decom-

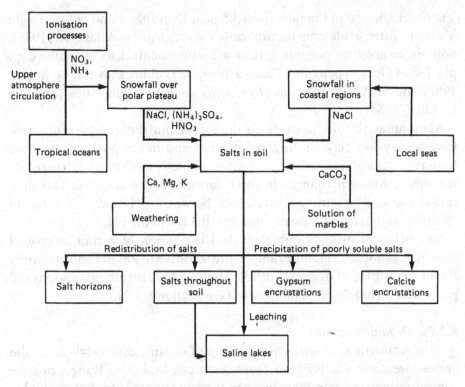

Fig. 8.4. Postulated origins of salts in antarctic soils. Redrawn from Claridge & Campbell (1977), with the additional input of ammonium and nitrate from ionisation processes in the atmosphere, as discussed by Zeller & Parker (1981).

position of ferromagnesian minerals in the dolerites and granites. Potassium was largely released from the weathering of micas while sodium was believed to be mostly of marine origin. The dissolution of marble and other calcareous rocks released carbonates while chloride, nitrate and sulphate were considered to be from a marine source. On maritime Signy Island a similar diversity of salt origins has been postulated: sodium and magnesium from the sea, potassium and some calcium from the chemical weathering of the rocks (aided by the freeze–thaw cycles which physically break down the rock), phosphate from the marine fauna, and nitrate and ammonia from the fauna and from rainfall (Allen, Grimshaw & Holdgate, 1967).

### 8.2.6   Soil pH
Acid as well as alkaline soils are represented throughout Antarctica, although the pH of soils with an extremely low moisture content is difficult to accurately assess. In a survey of 44 areas throughout southern

Victoria Land soil pH ranged from 3.5 near Don Juan Pond in the Wright
Valley to 9.0 in sands from the adjacent Victoria Valley (Atlas *et al.*, 1978).
Soils influenced by penguin guano are often neutral to alkaline; e.g.,
pH 7–8 in Gentoo penguin (*Pygoscelis papua*) colony soils (H. G. Smith,
1982) and in Adélie penguin (*Pygoscelis adeliae*) colony soils (Speir &
Cowling, 1984).

Many antarctic soils, particularly the continental desert representatives,
have a very low clay and organic content and hence possess very little
buffering capacity (Claridge & Campbell, 1977). The pH is therefore
responsive to slight changes in the balance between acidic and alkaline
salts. There is a strong correlation between pH and the ratio of
chloride : sulphate – the lower this ratio, the lower the pH.

Soil pH may strongly influence the distribution of certain microbial
groups. Cyanobacteria, for example, are sensitive to low pH and are rarely
found below pH 5 (Fogg *et al.*, 1973). High acidity also inhibits various soil
processes, including methanogenesis (see Section 8.4.2).

### 8.2.7    *Organic content*

Antarctic soils span the extremes of organic carbon levels. In the
ancient weathered soils of the Trans-Antarctic Mountain Range, organic
concentrations are extremely low and the opportunities for heterotrophic
nutrition are minimal. At the other extreme, the ornithogenic soils found
in the penguin colonies contain abundant levels of organic carbon and
provide a rich source of substrates for microbial heterotrophy. At Cape
Bird, Ross Island, the soils contained 21–24% organic carbon (dry weight),
with up to 61% weight loss on ignition. Old, abandoned penguin nesting
sites had much lower levels of soil organic-C; 6.2% or less (Speir &
Cowling, 1984).

The Dry Valley region contains marine and lacustrine (lake-derived)
sediments with a low but detectable content of organic carbon. The total
organic carbon (TOC) levels in a Taylor Valley sediment core to 313 m
varied from 0.02 to 0.31%. For the top 100 m of the core, the highest
values of TOC (up to 0.3%) as well as highest concentrations of plant
(phytoclast), microalgal (protistoclast) and fungal (scleratoclast) remains
occurred in the near-surface samples (Wrenn & Beckman, 1981).

In the maritime zone mosses and lichens release large quantities of
dissolved and particulate organic matter into the soil. Their tissues contain
high concentrations of sugars and sugar alcohols (polyols) which act as
compatible solutes, protecting the cells during periods of water stress
imposed by freezing (see Chapter 10). On Signy Island the lichens at a

*fellfield* (scree soils with a discontinuous cover of lichens and bryophytes) contained about one-quarter of their cellular dry weight as polyols throughout the year and the adjacent soils contained up to 1.5% dry weight as these compounds (P. V. Tearle, cited in Wynn-Williams, 1987). These polyols are preferred substrates for heterotrophs that in turn may precondition the environment for the phototrophs. In a colonisation experiment on the Signy Island fellfields there was a significantly greater final abundance of cyanobacteria in the soils enriched with moss extract (containing mostly the polyols arabitol and ribitol) than in soils enriched with glucose (Wynn-Williams, 1987). The polyols are released from the mosses and lichens during freeze–thaw cycles and there can be a rapid response by heterotrophs such as yeasts and bacteria to these pulses of organic carbon availability (Ellis-Evans & Wynn-Williams, 1985; Wynn-Williams, 1985).

### 8.2.8   *Nitrogen content*

The total nitrogen of antarctic soils is often unusually high. In the soils away from penguin rookeries this is primarily due to nitrates which, like other salts, have been attributed to various marine and atmospheric sources. Isotopic analysis of the $NO_3$ in Dry Valley soils, however, did not support a marine origin (Wada, Shibata & Torii, 1981). The $\delta^{15}N$ (ratio of ($^{15}N - {}^{14}N$) : $^{14}N$) was extremely low ($-11$–$-24$‰) by comparison with seawater (7‰) and typical soils (3–6‰). The authors interpreted these data as support for an atmospheric origin of nitrate derived from photochemical reactions and auroral activity.

The organic nitrogen of Dry Valley soils is low, 0.3–6.4 mmol N $g^{-1}$ dry soil, which is one to two orders of magnitude below the nitrate concentrations (Wada *et al.*, 1981). Very much higher concentrations are found in the ornithogenic soils. A single extraction of soil from the Cape Bird rookery yielded a remarkable 18% dry weight of soils as uric acid, but the decomposition product of this compound, ammonium, is often also present in high concentration. Nitrate levels in these Cape Bird soils were generally three orders of magnitude below the ammonium concentrations (Speir & Cowling, 1984). The $\delta^{15}N$ in these soils was high, 28.6‰ for the organic nitrogen, probably reflecting the marine origin of the penguin food, since the relative abundance of $\delta^{15}N$ tends to increase along a food chain, typically from 2.5‰ in marine plankton through to 15–20‰ in fish (Wada *et al.*, 1981).

The surface horizons of such soils may contain high levels of certain enzymes: alkaline phosphatase (cleaves organic-P to inorganic-P), protease (protein to amines), urease (urea to ammonia) and probably also uricase (uric acid to urea). At Cape Bird on Ross Island the potential activities of these enzymes were high enough to suggest that a large percentage of organic nitrogen and phosphorus could be broken down to inorganic forms each season (Speir & Ross, 1984). Actual rates of enzyme activity *in situ* were possibly limited by moisture availability – only five 'wet days' in which soil temperatures stayed above 0 °C for most of the 24 h cycle, were estimated to occur each year (Orchard & Corderoy, 1983). Recent guano contained high levels of dehydrogenase, indicative of actively metabolising bacteria. The gut microflora contained in fresh penguin excreta, rather than soil micro-organisms, was probably the dominant source of enzymes in the Cape Bird soils, although these proteins may be subsequently stabilised in abiotic soil complexes (Speir & Ross, 1984).

### 8.3    Microbial community structure

#### 8.3.1    *Phototrophic communities*

The distribution of edaphic algae and cyanobacteria has been examined in detail at many sites around the margins of Antarctica, including the Antarctic Peninsula (Broady, 1979d), South Orkney Islands (Broady, 1979e), Vestfold Hills (Broady, 1986), Mawson Rock (Broady, 1983), Mt Erebus (Broady, 1984b), the Dry Valleys (e.g. Broady, 1982a), coastal sites in the McMurdo Sound region (e.g., Broady, 1982a); and further north up the coast of Victoria Land, including Mt Melbourne (Broady, *et al.*, 1987). These wide-ranging surveys by a single investigator are especially valuable given the consistency of methods and taxonomic criteria. Most of the discussion which follows is based upon these observations.

The edaphic algal flora of Antarctica is not diverse. Less than 70 genera have been recorded in the soils throughout the region, and of these only a few species are common: the $N_2$-fixing cyanobacterium (cyanophyte) *Nostoc commune*, members of the cyanobacterial family Oscillatoriaceae (Fig. 8.5), the pennate diatom *Navicula muticopsis*, and the foliose chlorophyte *Prasiola crispa*. Other genera that are often recorded in temperate or tropical soils are conspicuously absent: for instance *Cylindrospermum*, *Protosiphon*, *Euglena*, *Botrydium* and many of the common chlorophytes of the Chlorococcales, Chlorosarcinales and Chaetophorales (see Broady, 1986, 1987).

There appear to be regional differences in species diversity. Terrestrial mosses throughout the world typically harbour a wide range of algal

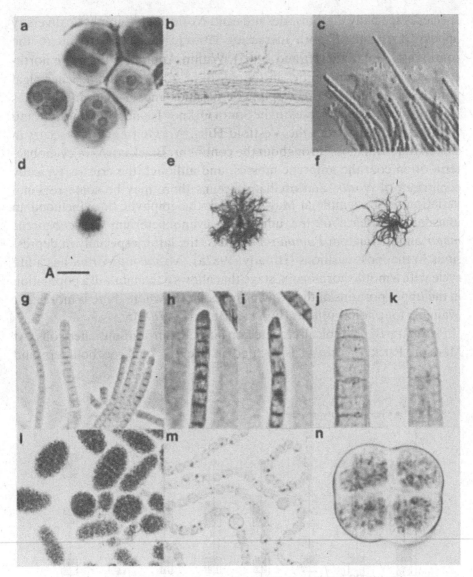

Fig. 8.5. Some terrestrial algae from the Vestfold Hills area. (*a*) *Gloeocapsa ralfsiana*; (*b*) *Microcoleus vaginatus*; (*c*) *Microcoleus chthonoplastes*; (d–f) oscillatoriacean cyanobacteria growing on agar plates; (g–k) oscillatoriacean trichomes; (*l*) microcolonies of *Nostoc*; (*m*) mature *Nostoc* (the large spherical cell is a heterocyst); (*n*) cubical aggregate of *Prasiococcus calcarius* cells. The scale bar (A) represents 10 $\mu$m in (*a*), (g–k) and (*n*); 30 $\mu$m in (*c*) and (*l*); 100 $\mu$m in (*b*); and 1 cm in (d–f). Reproduced from Broady (1985) by permission of the author and Academic Press, Australia.

species, especially diatoms, desmids and *Nostoc*. This epiphytic diversity appears to drop off with increasing latitude from the tropics to the continental Antarctic (Hirano, 1965). Within Antarctica the more northerly bryophyte communities in the maritime zone contain a richer algal flora (Broady, 1986). Eight epiphytic desmids and 17 diatom epiphytes were recovered from mosses in the South Orkney Islands, but only one and four, respectively, from the Vestfold Hills. *Nostoc* frequently occurs in association with moss throughout the continent. Black crusts of cyanobacteria often coat the antarctic mosses, and although the crust is typically comprised of *Nostoc* and oscillatoriaceans there may be some regional variation – for example at Mawson Rock the epiphytic flora included an unusual abundance of the unicellular cyanobacterium *Synechococcus major* and the diatom *Pinnularia borealis*, the latter especially in depressions in the moss cushions (Broady, 1982a). *Nostoc muscorum* has a life cycle with a motile hormogone stage that allows it to maintain a population in the upper portions of the growing moss shoots where there is more light available for photosynthesis (Fig. 8.6; Broady, 1979a).

*Prasiola crispa* is typically located on flushed, nutrient-enriched soils. At Mawson Rock, *P. crispa* is located in the lee of large boulders and

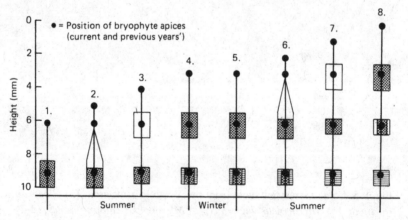

Fig. 8.6. Growth and life cycle model for *Nostoc muscorum* on moss shoots on Signy Island. At the beginning of the summer heterocystous thalli (shaded) are located near the previous year's shoot apices (1). Soon after the seasonal thaw the trichomes fragment and motile hormogones are released (unshaded); these move up the growing moss shoot (2). The hormogones reach the site of the old terminal bud (3) and there produce heterocystous thalli (4). With increased shading by the bryophytes the lower *Nostoc* communities become moribund (lined shading) while the motile phase of the cycle is repeated the next season (5–8). Redrawn from Broady (1979a).

hummocks where snow drifts accumulate, and where moulting Adélie penguins congregate towards the end of summer, ensuring a rich source of nutrients (Broady, 1982b). At sites on the coastal margin of the Vestfold Hills that receive both snowmelt and nutrient input from penguins, *P. crispa* forms vivid green stands, with associated populations of *Navicula muticopsis* and Oscillatoriaceae. In the organically rich, moist but infrequently flushed areas within the rookeries the algal growths are less prolific and more diverse; *Prasiococcus*, *Chlorella*, *Stichococcus*, *Bracteacoccus* and *Chlorococcum* as well as *N. muticopsis* and oscillatoriaceans commonly occur. Inland, the enriched soils around skua and snow petrel nests often contain green and dark blue-green crusts of Oscillatoriaceae and various chlorophytes. *P. crispa* is an infrequent resident of these sites, sometimes present in its immature, uniseriate filamentous stage.

Similar communities occur at coastal sites on Ross Island and along the margin of Victoria Land (Broady, 1985). Rich green algal crusts are common in areas with a high salt content and nutrient enrichment by birds. These usually consist of *Prasiococcus calcarius* (Fig. 8.5) and sometimes the small branching thalli of a chaetophoralean similar to *Coccobotrys*. *Prasiola crispa* occurs on water-flushed, guano-enriched sites as green foliose thalli, while a second species, *P. calophylla* occurs outside the bird-influenced areas, generally attached to stones between which water percolates.

In the mineral soils flushed with water but with little or no animal influence the dominant species tend to be cyanobacteria, including *Nostoc, Scytonema* and *Calothrix*. *Nostoc commune* forms leathery, irregularly-lobed colonies and may contain fungal hyphae to a variable extent. It appears to become increasingly lichenised with decreased moisture availability and the fully lichenised material is known by the name *Collema*. *Prasiola crispa* may also appear in its lichenised form (*Mastodia tesselata*) under such conditions in the maritime zone (P. A. Broady, pers. comm.).

Members of the Oscillatoriaceae are especially common on moist to wet mineral soils. This group is taxonomically difficult but antarctic specimens can be assigned to the classical genera *Oscillatoria, Lyngbya, Phormidium* and *Schizothrix*. *Microcoleus* (especially *M. vaginatus* and *M. chthonoplastes*) may also be locally abundant. This and the other filamentous genera have an important influence on the physical properties of the soil by binding together the particles with their intertwining filaments and mucilaginous sheaths. In an artificial substrate experiment on the Signy Island fellfields various oscillatoriacean species played different roles in the colonisation process (Wynn-Williams, 1987). A large-sized *Phor-*

*midium* (*c.* 5–7 $\mu$m $\times$ 50 $\mu$m) colonised the coarse-grained substrates and appeared to be important in stabilising soil crusts. Colonisation was accelerated, however, by the attachment of 'microbial rafts' containing various sized oscillatoriacean trichomes as well as algae and bacteria bound together with mucopolysaccharide-like substances ('mucigels'). These microbial aggregates were released from developed communities in the surrounding catchment.

In moist but unflushed sites, oscillatoriacean species may stain the soil a distinctive red-brown or blue-green colour. These communities typically contain a small associated population of chroococcalean cyanobacteria, diatoms and unicellular chlorophytes, and are the first microbial communities to colonise moraines emerging at the edge of the polar ice sheet (Broady, 1986).

The geothermal habitats of Antarctica contain a distinctive and unusual flora. Microalgae were first observed on the warm slopes of Mt Erebus as red-coloured felts on soil surfaces at 43 °C (Lyon & Giggenbach, 1974). These communities have been subsequently found to contain a surprising diversity of microalgae (Broady, 1984b) including four cyanobacterial species and 11 unicellular chlorophytes. The thermal communities show clear zonation according to soil surface temperature. (Fig. 8.7).

In the hottest ground, at 40–60 °C, the dark grey to blue-black felts were dominated by an unidentified cyanobacterium similar to *Lyngbya*. In the adjacent cooler areas at 30–40 °C, *Phormidium fragile* formed red-brown felts. Between 20–30 °C, *P. fragile* and unicellular chlorophytes occurred amongst abundant moss protonema, and at lower temperatures to 10 °C the soil was covered by green crusts of the chlorophyte unicells. At other warm-ground sites the *Lyngbya*-like cyanobacterium was absent and the dominant was the cyanobacterium *Mastigocladus laminosus*. This species is capable of heterocyst production for $N_2$-fixation, and is a common constituent of the flora of geothermal springs in other parts of the world.

A more diverse algal flora occurs in the steaming ground at the summit of Mt Melbourne (Broady *et al.*, 1987). As on Mt Erebus these soils contained the cosmopolitan thermophile *Mastigocladus laminosus, Phormidium fragile*, and a range of the unicellular chlorophytes, but additional microalgal inhabitants included mucilaginous colonies of *Gloeocapsa magma*, and filaments of *Stigonema ocellatum* and *Tolypothrix bouteillei*. The thermophilic *Lyngbya*-like alga found on Mt Erebus was not detected on Mt Melbourne. *M. laminosus* grew at the higher end of the temperature range in soils from 25–40 °C. The other cyanobacteria and chlorophytes only occurred at temperatures below 25 °C.

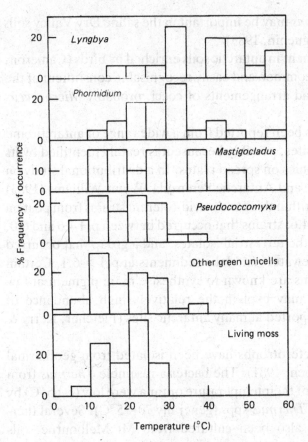

Fig. 8.7. The zonation of phototrophs on Mount Erebus in relation to soil surface temperature. Redrawn from Broady (1984b).

### 8.3.2   *Bacterial heterotrophs*
*Population structure*

A wide range of morphological types of bacteria have been observed at many sites in Antarctica, but most of the studies to date have relied upon enrichment culture techniques that may select for unusual or metabolically unimportant species in the overall community. *Pseudomonas*, *Achromobacter* and *Alcaligenes*, all gram-negative bacilli, have been isolated from the Antarctic Peninsula region (Block, 1984). *Corynebacterium*, *Arthrobacter* and *Micrococcus* have been isolated from Dry Valley soils (Cameron, Morelli & Johnson, 1972) and a similar assemblage has been recorded in the volcanic ash-based soils of Deception Island in the maritime Antarctic (Cameron & Benoit, 1970). A halotolerant species of *Planococcus* may be important in the saline Dry Valley soils (Miller, Leschine & Huguenin, 1983).

*Micrococcus* is prominent in antarctic soils enriched by birds (Cameron, 1971). At Cape Bird, a common and easily recognisable constituent of the soil microflora was tetrad arrangements of cocci, probably *Micrococcus* (Ramsay, 1983).

*Chromobacterium* has been reported from a wide range of antarctic and subantarctic terrestrial sites, possibly because it is readily identified by its conspicuous purple colonies on spread plates. In a distributional study on islands of the Scotia Arc and Antarctic Peninsula, Wynn-Williams (1983) identified three groups within this genus: acid-tolerant strains from peat on South Georgia at pH < 4.6; strains that occurred between pH 4.6 and 6.0, which included most of the terrestrial isolates; and a group that occurred in mineral material, lakewaters and lake sediments at pH > 6.1. Certain coloured micro-organisms are known to synthesise more pigment at low temperatures, and this may explain the relatively high abundance of chromogenic bacteria reported at many antarctic sites (Fletcher, Kerry & Weste, 1985).

Thermophilic microheterotrophs have been isolated from geothermal sites on Mt Erebus (Green, 1981). The bacteria resemble *Thermus* from temperate hot springs, but their temperature optima were low (*c.* 65 °C) by comparison with typical *Thermus* spp. (generally 70–75 °C). Several thermophilic bacteria have also been cultured from Mt Melbourne soils (Broady *et al.*, 1987).

Soil from a moss site at Cape Bird yielded a diverse assemblage of aerobic bacteria, including members of the genera *Bacillus*, *Achromobacter*, *Arthrobacter*, *Aerobacter*, *Pseudomonas* and various flavobacteria (Greenfield & Wilson, 1981) as well as the actinomycete *Streptomyces*.

### Biomass structure

Bacterial cell densities are especially high in the organically rich ornithogenic soils at penguin nesting sites. At Cape Bird the concentration of bacteria averaged about $2 \times 10^{10}$ cells $g^{-1}$ soil (dry weight), and about $0.1 \times 10^{10}$ $g^{-1}$ at a site outside the rookery (Ramsay, 1983). The bacteria ranged from very small cocci, 0.25 $\mu$m in diameter, to large cocci and rods up to $1 \times 3$ $\mu$m. Cell volumes were converted to dry weights by assuming standard relationships, and these yielded total biomass estimates ranging from $0.04 \pm 0.01$ mg $g^{-1}$ soil at the unoccupied site, to 5.1 $\pm$ 0.8 mg $g^{-1}$ at the occupied nesting area. There was a significant correlation between soil moisture content and bacterial concentrations, but not biomass.

Much lower, yet detectable levels of microbial biomass have been recorded in inland continental soils. Friedmann, La Rock & Brunson (1980) analysed three high altitude Dry Valley soils from sites near sandstone boulders that contained a cryptoendolithic microflora (see Chapter 9). These soils contained unusual levels of organic nitrogen ($1.97–6.06$ g m$^{-2}$), no chlorophyll *a*, but higher levels of ATP than in the adjacent microbially-colonised rocks: $19.4–46.3$ mg ATP m$^{-2}$ compared with $0.76–18.4$ mg m$^{-2}$ in the rocks. This material was possibly derived from the endoliths growing on the adjacent rocks. These organisms promote exfoliating weathering of the rocks in which layers of the rock crust peel off dispersing microbial cells to the ground. The authors suggested that the chlorophyll *a* then decomposes rapidly while the ATP is preserved in the frozen soils. It remains possible, however, that some of the ATP was associated with metabolically active or viable heterotrophs in these soils.

### 8.3.3 Fungal communities
#### Population structure
Antarctic soils support a wide range of fungal residents including basidiomycetes, oomycetes and ascomycetes. This latter group is especially important and includes yeasts that sometimes dominate the soil microflora, as well as species that may be endemic to Antarctica.

In a seven year survey by Atlas *et al.* (1978) in southern Victoria Land, 25% of all soils examined contained yeasts in sufficient quantities to be enumerated. The dominant genera were *Cryptococcus*, *Aureobasidium* (a 'yeast-like' fungus), *Candida*, *Torulopsis*, *Rhodotorula* and *Sporobolomyces*. The most frequently isolated yeast was *Cryptococcus albidus*, which the authors were also able to isolate from temperate desert soils and dry Alaskan soil. Two other species, *C. luteolus* and *C. laurentii* were more frequently isolated from antarctic soils of higher moisture content and were also constituents of Alaskan tundra soils. *Aureobasidium pullulans* was also isolated from a wide variety of habitats including sandy soils of the Dry Valleys, and gravelly sand from the Strand Moraines near a skua nesting site, McMurdo Sound.

In an earlier comparative study of soils from the Ross Dependency, and from a site in Greenland (Mestersvig, 72°N) three genera were common to samples from both regions (*Cryptococcus*, *Candida* and *Rhotorula*), while a fourth genus, *Debaryomyces* (sexual form of *Torulopsis*), was only recovered from the antarctic samples (Di Menna, 1966). *Cryptococcus laurentii* and *C. luteolus* were especially frequent isolates from both polar regions. Most of the samples that contained yeasts also contained moss, lichen or microalgal material, but there was no correlation between the abundance of yeasts and the presence of penguins or skuas. Soil pH varied over a similarly wide range (4.5–9.7) to that reported by Atlas *et al.* (1978), and showed no correlation with yeast species composition or abundance.

All of these studies depend upon enrichment assays in artificial media that may be unsuitable for, or even inhibitory to, some components of the native microflora. An alternative culture approach with Dry Valley soils has yielded a variety of additional yeasts, including certain endemic species. Vishniac (1983) employed a dilute nutrient medium in which the fungi were allowed to grow out (become enate) from the soil particles. Two yeasts were isolated by this method, *Trichosporon cutaneum* and an apparently indigenous species subsequently named *Cryptococcus vischniacii* var. *associalis*. Additional micro-organisms identified in the soils by this technique were the zygomycete *Mucor racemosus*, dermatiaceous and foliaceous hyphomycetes, and the microalgae *Chlorosarcina* and *Trebouxia*.

Coastal sites in the Antarctic contain a diverse assemblage of fungi. At a moss site at Cape Bird, Ross Island, members of the following genera have been isolated: *Epicoccum, Aureobasidium, Cladosporium, Alternaria, Penicillium, Aspergillus, Phialophora, Mortierella, Trichoderma, Mucor* and *Chrysosporium* (Greenfield & Wilson, 1981).

Filamentous fungi were widespread through the ice-free sites of Mac. Robertson and Enderby Lands, but keratinophilic forms were apparently absent (Fletcher *et al.*, 1985). Fungi were always present in organic soils, but were absent from penguin guano possibly because of the microbial inhibitor acrylic acid (Boyd, 1967). Like other sites in Antarctica the species diversity was limited. The most frequently encountered isolates were *Cladosporium herbarum* and *Penicillium* spp., especially *P. brevicompactum* and *P. cycloporium*. The ascomycete *Thelebolus microsporus* was isolated from two sites in Mac. Robertson Land, and had been previously reported from the Bunger Hills, Balleny Islands, and Vestfold Hills indicating a widespread distribution in the eastern half of Antarctica. About half of the isolates were sterile (no sexual stages), which seems to be a common feature of polar fungi.

Mummified seals provide an unusual fungal habitat in the McMurdo Sound region (Greenfield, 1981). These animals (generally crabeater seals, *Lobodon corcinophagus*), wander inland and eventually die, but are preserved by desiccation. Three fungal species have been isolated from the colonies that are visibly conspicuous on these seals: *Phialophora fastigiata*,

*P. gopugerotii* and *P. dermatitidis*. The latter two species are pathogenic to man causing skin lesions.

In the fumarole sites on Mt Erebus, where soil temperatures typically exceed +20 °C various soil fungi form thick wefts of mycelial growth. *Penicillium*, *Aspergillus* and *Neurospora* were recorded as the fungal dominants in the first biological survey of the region (Ugolini & Starkey, 1966). A less diverse assemblage of thermotolerant but not thermophilic fungi and actinomycetes have been isolated from hot soils at the summit of Mt Melbourne (Broady *et al.*, 1987). The area contained patches of bryophytes, identified as *Campylopus pyriformis* (a moss) and *Cephaloziella exiliflora* (a liverwort), that have given rise to small peat-like deposits. These organically-rich soils contain visible wefts of the thermotolerant/thermophilic organisms *Thermomonospora* sp., *Chaetomium* sp., *Malbranchea pulchella* var. *sulfurea* and *Myceliophthora thermophile* and the actinomycete *Streptomyces coelicolor*.

### Biomass structure

The population density of the soil yeasts varies enormously with location, from one colony-forming unit (presumably one yeast cell) in the Wheeler Valley (a Dry Valley) to near $5 \times 10^3$ g$^{-1}$ soil from the Strand Moraines (Atlas *et al.* 1978). In soils influenced by human activities in the vicinity of McMurdo Station the yeast count rose by three orders of magnitude to $5 \times 10^5$ g$^{-1}$. Yeasts were the only detectable fungi in many of the soils, but often comprised a small percentage of the total number of heterotrophs that formed colonies in the enrichment assay. In the McMurdo Station soils the yeasts totalled 10% of the total heterotrophs, but at the Strand Moraines and at Wheeler Valley they made up only 0.5 and 0.001%, respectively, of the total heterotrophic population.

Vishniac (1983) noted that quantitative data from the enation technique was premature, but concluded 'that yeasts peculiar to the Antarctic dominate in Dry Valley soils and at least two microcolonies of yeasts may be present in a gram of soil in relatively sheltered, productive sites' but given the deficiencies of the method it is possible that 'the total biomass of yeasts in these inhospitable soils rivals that of the endolithic communities'.

### 8.3.4 Protozoan communities

Protozoa are an important microheterotrophic component of moss-dominated ecosystems in the maritime Antarctic (Davis, 1980). Culture assays of a range of fellfield soils on Signy Island yielded 21 species of protozoa: 11 flagellates, 9 ciliates and a testate rhizopod. The species

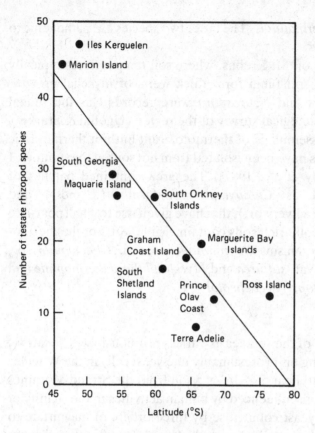

Fig. 8.8. The decrease in species diversity of testate amoebae in antarctic soils with increasing latitude south. Redrawn from Smith (1982).

diversity increased with increasing organic content of the soil. For the samples with low moisture and organic content there was also a positive correlation between number of species and pH (Smith, 1984). Order of magnitude higher biomass levels of protozoa occur in the peat-forming bryophyte habitats on Signy Island (Smith, 1978).

A diverse assemblage of protozoa has also been recorded in the moss sites at Cape Bird on Ross Island. The dominant species was a ciliate, probably *Cyclogramma membranella* (Greenfield & Wilson, 1981). In general, however, protozoan species diversity seems to decrease with increasing latitude south (Fig. 8.8).

A single protozoan species has been recorded in the warm soils at the top of Mt Melbourne; the testate amoeba, *Corythion dubium*. Neither this nor other protozoan species have been detected in the Mt Erebus thermal soils (Broady, 1985).

### 8.4 Microbial processes

#### 8.4.1 Photosynthetic production

Photosynthetic studies to date have been restricted to populations of the foliose alga *Prasiola crispa* and the cyanobacterium *Nostoc commune*. Both species in southern Victoria Land showed photosynthetic activity to well below 0 °C; *P. crispa* to −15 °C and *N. commune* to −5 °C (Becker, 1982). *Nostoc* remains viable after freezing to −70 °C, and seems highly tolerant of freeze–thaw cycles (see section 10.4).

#### 8.4.2 Nitrogenase activity

The nitrogen-fixing cyanobacterium, *Nostoc*, is a major contributor of soil nitrogen and organic carbon at many ice-free localities in Antarctica. Studies in the Vestfold Hills have shown that the seasonal dynamics of nitrogenase activity are closely regulated by temperature and moisture availability (Davey & Marchant, 1983). Maximum rates were recorded in December and January when the *Nostoc commune* mats were well supplied with meltwater and ground temperatures were up to 8–10 °C (Fig. 8.9). By late February the soil temperatures were constantly below zero and the meltwaters had evaporated, but low levels of nitrogenase activity persisted until early April when temperatures fell below −7 °C.

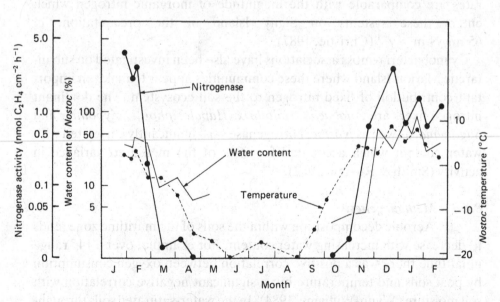

Fig. 8.9. Seasonal changes in nitrogenase activity, moisture content and temperature of a terrestrial population of *Nostoc commune* in the Vestfold Hills. Modified from Davey & Marchant (1983).

The *Nostoc* mats recovered their ability to fix nitrogen in early October when surface temperatures rose to $-7.6\,°C$, but the greatest increase in nitrogenase activity was during November when the soil temperatures increased above $0\,°C$ and the sites became inundated by melting snow.

*Nostoc commune* in Antarctica often occurs in association with bryophytes. In the Vestfold Hills the highest nitrogenase activities were recorded in a moss–*Nostoc* flush area where the bryophytes included *Bryum antarcticum. B. algens, Grimmia antarctici* and *Sarconeurum glaciale* (Davey 1986). Nitrogen fixation rates were estimated indirectly using the artificial substrate acetylene which is converted by the nitrogenase enzyme system to ethylene. The ratio between actual $N_2$-fixation and measured ethylene production can vary over a wide range under some circumstances, but assuming the theoretical ratio of three moles of ethylene produced per mole of $N_2$ fixed the estimated annual rates of nitrogen fixation were 119 mg N m$^{-2}$ for the damp moss–*Nostoc* association, 10 mg N m$^{-2}$ for moss–*Nostoc* cushions at a more exposed and drier site, and 52 mg N m$^{-2}$ for the free-living mats of *Nostoc* (Davey & Marchant, 1983). These values lie well above Horne's (1972) estimate of 1.8 mg N m$^{-2}$ for *Nostoc commune* on Signy Island. More recent estimates for the moss–cyanobacteria communities on Signy Island have given $N_2$-fixation values of 46 (dry turf) – 192 (wet carpet) mg N m$^{-2}$ y$^{-1}$. These rates are comparable with the magnitude of inorganic nitrogen which enters these systems on Signy Island in the precipitation, *c.* 65 mg N m$^{-2}$ y$^{-1}$ (Christie, 1987).

Cyanobacteria–moss associations have also been investigated on subantarctic Marion Island where these communities appear to make an important contribution of fixed nitrogen to the soil ecosystem. The dominant nitrogen fixers are *Anabaena, Calothrix, Hapalosiphon, Tolypothrix* and *Stigonema*, as well as *Nostoc*. Nitrogenase was significantly correlated with water content which accounted for 69% of the measured variance in activity (Smith & Russell, 1982).

### 8.4.3   *Methanogenesis*

Aerobic decomposition within the soils of the maritime zone tends to decrease with increasing water content. For example, over a $14\,°$ range in latitude there was a positive correlation between oxygen consumption by peat soils and temperature but a significant negative correlation with soil moisture (Wynn-Williams, 1984). In the water-saturated soils throughout this zone anaerobic decomposition of organic material to methane probably accounts for a considerable portion of the total carbon flux (Wynn-Williams, 1984).

In a wet moss site at Signy Island the rate of methane production over the summer averaged 1.24 mg C m$^{-2}$ d$^{-1}$ (Yarrington & Wynn-Williams, 1985). This value seems extremely low by comparison with earlier measurements of total $CO_2$ release by these peat soils (e.g. about 600 mg C m$^{-2}$ d$^{-1}$ for the wet peat on Signy Island, Wynn-Williams, 1984), but it is possible that a large amount of the methane produced is oxidised to $CO_2$ by methylotrophs in the upper aerobic parts of the soil profile. Net methane production at Signy Island was stimulated by increasing pH and temperature (Fig. 8.10) indicating that the low ambient values

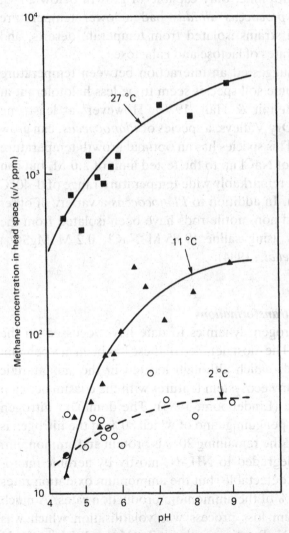

Fig. 8.10. Methanogenesis in the soil beneath a wet moss community on Signy Island and its relationship to temperature and pH. Redrawn from Yarrington & Wynn-Williams (1985).

of both variables limited the rates of methanogenesis in these soils (Yarrington & Wynn-Williams, 1985).

### 8.4.4    *Temperature–salinity adaptation*

Many of the fungal isolates from antarctic soils appear to grow over a wide temperature range. Atlas *et al.* (1978) report that most of their isolated soil yeasts were psychrotrophic, capable of growth at both cold (5 °C) and warm (20 °C) temperatures. *Candida scottii* and some *Crypto-coccus* strains were psychrophilic, only capable of growth below 20 °C. Antarctic isolates of *Cryptococcus albidus* had a lower temperature optimum for growth than strains isolated from temperate deserts, and unusually slow utilisation rates of lactose and galactose.

Various authors have suggested an interaction between temperature and salt tolerance since some soil species seem to be less halotolerant at low temperatures (e.g. Benoit & Hall, 1970). However, at least one bacterial isolate from the Dry Valleys, a species of *Planococcus*, can grow in cold saline conditions. This species has an optimal growth temperature of 25 °C at concentrations of NaCl up to the tested limit of 2.0 M, but can grow at this salinity over a remarkably wide temperature range of 0–40 °C (Miller & Leschine, 1984). In addition to *Planococcus* a variety of other cocci as well as motile and non-motile rods have been isolated from the soils of the Dry Valleys using saline (0.85 M NaCl, 0.2 M $MgSO_4$) enrichment media (Miller *et al.*, 1984).

### 8.4.5    *Microbial nitrogen transformations*

Studies of soil nitrogen dynamics to date have focussed on the coastal ornithogenic soils. The most detailed of these has been in a penguin rookery on Marion Island, which although located in the subantarctic (46 °S) probably shares many ecosystem features with the coastal penguin rookeries in the Antarctic (Lindeboom, 1984). The dominant nitrogen source in this ecosystem is penguin guano of which 80% of the nitrogen is in the form of uric acid and the remaining 20% is protein and ammonium. The uric acid is rapidly degraded to $NH_4$-N, mostly by aerobic micro-organisms. Nitrifiers were detectable, but the ammonium oxidation rates were slow, only about 2.5% of the ammonium production rates. A much more important ammonium loss process was volatilisation which was estimated from changing N:P ratios in the soil. Of the 275 kg $NH_4$-N introduced each day to the rookery some 220 kg was volatilised.

### 8.4.6  *Bacterial and fungal heterotrophy*

The proportion of metabolically active cells in antarctic soils seems extremely low. At Cape Bird, less than 8% of the bacteria in the ornithogenic and nearby soils took up radio-isotopically labelled glucose (Ramsay, 1983). A wide array of heterotrophic microbial species have been isolated from antarctic soils but it remains unknown whether many of these are metabolically active in this cold, dry terrestrial environment.

### 8.5  **Trophic structure**

As in the freshwater habitats of this region larger organisms that play important functional roles in temperate latitude ecosystems are absent from antarctic soils, and the pattern of carbon and energy flow is correspondingly reduced in complexity. Higher plants as well as earthworms, land molluscs, higher insects (Diptera, Coleoptera and Hymenoptera) and spiders are only to be found in the most northerly parts of the maritime region and in the subantarctic. Similarly the terrestrial birdlife and microtine populations that are so characteristic of the arctic tundra biome are completely absent from most of Antarctica.

Cyanobacteria, algae, lichens and bryophytes are the only phototrophs throughout most of the region. These fuel the heterotrophic bacteria and fungi, although thick peat deposits in certain areas attest to the generally slow rates of decomposition. In some habitats the breakdown of phototrophic biomass proceeds by anaerobic as well as aerobic pathways of decomposition. For example, methanogenesis appears to be an important loss pathway for carbon in certain maritime moss communities.

Herbivory is relatively unimportant in most of the terrestrial Antarctic. Locally, however, there may be abundant populations of two herbivorous arthropod groups, springtails (Collembola) and mites (Acari). Rotifers, nematodes and tardigrades are to be found in many of the soils, but are most active in association with the moist communities of moss or semi-aquatic algae.

Each of these grazers may have specialised food preferences. Collembolans at Signy Island (*Cryptopygus antarcticus*) and the Victoria Land coast (*Gomphiocephalus hodgsoni*) did not feed on the healthy bryophytes that they were found in association with, but instead grazed on fungal hyphae, algae, and dead and decaying bryophyte tissue (Fitzsimmons, 1971; Broady, 1979). Three species of mites on the Victoria Land coast fed on soil algae, and to a lesser extent fungi, while a fourth was predatory on the other soil arthropods (Fitzsimmons, 1971). Four groups of nematodes could be distinguished at Signy Island based on their feeding preferences: omnivores, predators, fungal feeders and microbial feeders, the latter species eating bacteria but also some microalgae and fungi (Spaull, 1973). The total biomass of all of these groups, however, is generally small relative to the microbiota. At the two moss sites at Signy Island, for example, protozoa accounted for more than 75% of the total invertebrate biomass, and more than 10% of the total heterotrophic respiration in the soil (Davis, 1981). By comparison the Rotifera, Tardigrada, Nematoda, Acari and Collembola collectively respired less than 0.5% of the total. This extremely small contribution by animals other than protozoa to the total carbon flux contrasts markedly with soil ecosystems in other parts of the world. For example, in arctic tundra the soil fauna (excluding protozoa) may account for up to several per cent of total heterotrophic respiration. The relatively low invertebrate activity in Signy Island soils may contribute to the slow decomposition rates that in combination with the slow, temperature-depressed rates of microbial heterotrophy allow the accumulation of large standing stocks of undecomposed organic material (Davis, 1981).

# 9

# Lithic ecosystems: the rock environments

## 9.1    Introduction

Antarctic microbes can be found in an enormous range of environments but perhaps their most unusual habitat is on or within rock. These *lithic* communities occur wherever there are exposed, translucent rocks such as marble, granite and sandstone. The assemblage commonly includes microalgae, cyanobacteria, fungi and heterotrophic bacteria with the phototrophs and fungi often associated symbiotically as a microscopic form of lichen. The community biomass is much lower than damp soil environments (see Chapter 8) but in certain parts of Antarctic these various rock-dwelling microbial species are the dominant life-forms.

Four major types of lithic habitat have been recognised: *epilithic*, the exposed rock surface; *chasmolithic*, fissures or cracks inside the rocks; *cryptoendolithic*, the internal airspaces between the rock crystals; and *sublithic* (or *hypolithic*), the undersurfaces of rocks, particularly translucent ones, embedded in the soil. Each of these communities experiences a distinct microenvironment that differs radically from the surrounding soil, snow and atmosphere. Each microhabitat has its own unique properties, and the community structure is correspondingly varied. Low temperatures and moisture may limit microbial growth in these environments, but in the cryptoendolithic and perhaps also chasmolithic habitats attenuated light and a shortage of physical space may further restrict the population size.

## 9.2    The environment

A combination of environmental extremes in the rock habitat limits microbial growth to certain rock types, and to only a brief period of activity each year. The three variables that appear to be pre-eminently important in controlling metabolic activity in these environments are light, moisture and temperature. The most detailed picture of the lithic habitat has been built up from a multidisciplinary study on Linnaeus Terrace, a high altitude site in the Asgard Range (1600 m ASL, 77 °S, 161 °E) in the McMurdo Dry Valleys region (also referred to as the Ross Desert, see Appendix 1). The porous rocks there are heavily colonised by cryptoendoliths, but the micro-organisms are only active when their narrow set of environmental requirements is met (Friedmann, McKay & Nienow, 1987). These include:

(i) Photosynthetically available radiation (PAR) at the surface of the rock above $100 \, \mu E \, m^{-2} \, s^{-1}$.

(ii) Relative humidity above 75% in the pore spaces.

(iii) Temperatures above $-10 \, °C$.

Continuous data for these environmental parameters were transmitted by satellite from Linnaeus Terrace from 1984 to 1986 and showed that the period of cryptoendolithic growth (as defined by these criteria) began no earlier than mid-November and extended to no later than early March. In this habitat the microbial community was active for less than $1000 \, h \, y^{-1}$.

### 9.2.1    Rock type

The geology of each antarctic region exerts an overall influence on the type of microbial community that may inhabit the rocks. The chemical properties of the rock appear to be relatively unimportant, but physical characteristics such as porosity, colour and transparency dictate its suitability as a microbial habitat. The micro-organisms do not appear to actively penetrate the rock by solubilisation, and therefore only rocks permeated with cracks or with a porous structure are colonised (Friedmann, 1982). In the Vestfold Hills area quartz stones, charnockite gneisses and other rocks were readily split or broken into flakes indicating the availability of habitats for chasmolithic growth. The phototrophic endoliths require light and these communities therefore only colonise translucent rocks. In the McMurdo Dry Valleys Beacon sandstone (orthoquartzite), granite, granodiorite and Koettlitz marble provide suitable habitats with adequate space between the crystals for microbial growth (Fig. 9.1). In these rock types sufficient light for photosynthesis can penetrate the rock to several millimetres depth. Dark and non-porous volcanic rocks such as Ferrar dolerite are rarely colonised.

The permanence of a rock surface also dictates its suitability for microbial growth. The windward side of rocks may be subject to maximum erosion which prevents the establishment of lithic communities. In the Vestfold Hills, for example, chasmoliths were usually found exclusively on the protected leeward rock faces (Broady, 1981b), but in other areas this

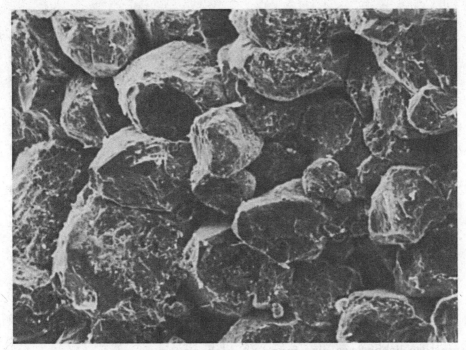

Fig. 9.1. The cryptoendolithic habitat. Scanning electron micrograph of the sandstone rock colonised by cyanobacteria, chlorophytes and microscopic lichens in the Dry Valleys region.

pattern is modified by temperature effects (see Section 9.2.4). In the Dry Valley sandstones a moderately silicified surface layer seems to be required for continuing microbial colonisation and growth. This smooth layer stabilises the rock against grain-by-grain weathering (Fig. 9.1; see also Fig. 9.4a below) and enhances water retention. Unsilicified surfaces are not colonised, and neither are heavily silicified crusts, which appear to prevent water absorption (Friedmann & Weed, 1987).

Chasmoliths occupy two types of rock habitat: vertical fissures and horizontal cracks. The former are common in the quartz glacial debris that is widely distributed over the Vestfold Hills. The charnockite substrates in the Mawson Rock area are often exfoliated into thin flakes with the cracks developed more or less parallel to the surface. The chasmoliths develop in these air spaces beneath 1–6.5 mm of rock (Broady, 1981b). Chasmoliths are also commonly observed in the fine fracture zones passing through marble, e.g. at Jane Col on Signy Island (Wynn-Williams, 1987), and at Marble Point in the McMurdo region.

### 9.2.2   Moisture availability

Melting snow is believed to be the major source of water for many lithic communities, but the availability of meltwater is dictated by the microtopography of the rock surface (Friedmann, 1978; Broady, 1981b). Chasmoliths occur in rocks that trap snow and ice, and are especially abundant in the lee sides of boulders and hills, in rock depressions, and in coastal regions covered by the ice foot (Broady, 1981). The porous rocks which contain cryptoendoliths readily absorb water from melting snow and this water may remain stored in the rock for considerable periods of time either as liquid capillary water or as water vapour. For example, after a snowfall on Linnaeus Terrace the relative humidity within interstices of the rock remained above 80% for at least five days while it repeatedly dropped below 20% in the atmosphere. Like desert lichens elsewhere, the endoliths are likely to remain fully active at this level of humidity within the rock (Kappen, Friedmann & Garty, 1981).

Epilithic communities probably also take up water directly from the air. The relative humities of the Dry Valley region tend to be low, but probably not as low as hot desert regions of the world where in addition to endoliths crustose lichens are also found. A higher vapour pressure is probably maintained in the near-surface boundary layer than in the overlying atmosphere (Kappen *et al.*, 1981).

### 9.2.3   Light availability (PAR)

The interior of rock environments is characterised by a steep light gradient. For the Linnaeus Terrace community about 0.1% of ambient light reaches the top of the lichen zone immediately under the rock crust, and about 0.01% reaches the stratum of green algae (*Hemichloris*). However, the light scattering within this rock medium is greatly reduced when the spaces between the crystals are filled with water, and penetration of light is correspondingly greater. Wetting the rock increases the availability of PAR to 1% under the rock crust and 0.1% in the *Hemichloris* stratum (Friedmann & Ocampo-Friedmann, 1984). This effect is further evidence of the high porosity of the rock types colonised by endoliths.

### 9.2.4   Radiation and temperature

Lithic microbial populations in the Dry Valleys appear to be most abundant in the rock faces which receive the longest period of direct insolation, and thus especially in the north-facing or horizontal rock surfaces (Friedmann, 1978). Other features may lessen the advantages of north-facing slopes. At Linnaeus Terrace, a continuously monitored northeastern face was considerably warmer than a nearby horizontal surface, with temperatures above 0 °C for five times longer than the flat surface. However, the sloped surface was out of the direct sun each day, and therefore fell to lower temperatures. The continuous humidity data indicated that the horizontal surface remained moist for longer periods than the sloping surface (Friedmann *et al.*, 1987). South-facing surfaces rarely heat above 0 °C and are therefore little colonised by endoliths. However, this pattern is not seen at the Vestfold Hills because of the effects of wind. North-facing rocks are normally devoid of algae because of the sand-blasting effect of the prevailing northerly and the absence of moisture-providing snow drifts on this windward side of the rock (Broady, 1981c).

Rock materials have a relatively high heat capacity; for example sandstone has ten times the heat capacity of uncompacted snow (see Appendix 2). Rock can therefore store heat to maintain temperatures well above ambient. North-oriented rock faces exposed to the sun typically rise to temperatures 1–20 °C higher than in the surrounding air. Similarly warm temperatures have been measured over the diurnal cycle in sublithic habitats in the Vestfold Hills (Broady, 1981d).

The surfaces of the rocks may be exposed to rapid changes in temperature that in part are controlled by air movements. Locations in the Dry Valleys often experience gusty winds which at the near-zero temperatures in summer can induce a rapid sequence of freezing and thawing. For example, on a midsummer's day at the Linnaeus Terrace site a rock surface fluctuated between −1.8 and 5.9 °C over a period of just 42 min, moving across 0 °C no less than 14 times. Friedmann (1982) identifies this aspect of the environment as 'probably the most important single factor responsible for the abiotic rock surface'. By contrast, in the region of the cryptoendoliths three millimetres below the rock, surface temperatures remained about 0 °C at all times, ranging from 1.7 to 6.1 °C.

Two important time-scales of temperature variation have been clearly defined in the sandstones of Linnaeus Terrace (McKay & Friedmann, 1985). Sandstone has a relatively high thermal conductivity (about five times higher than dry sand, for example; see Appendix 2) and the fluctuations in temperature over the 24 h cycle therefore penetrate to considerable depth in the rock (Fig. 9.2). This diurnal temperature wave was seen down to at least 20 mm below the surface, well into and below the zone of cryptoendolithic micro-organisms. The temperatures in the rock reached 12 °C while air temperatures peaked at 0.6 °C. The diurnal temperature range in the rock was 15–20 °C whereas the diurnal air temperature

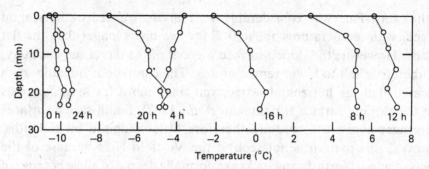

Fig. 9.2. Diurnal heating and cooling in the surface rock zone. This vertical sandstone rockface on Linnaeus Terrace was directly exposed to the sun, with maximum insolation at 0800 h. Cryptoendolithic lichens were located in the depth region 2–10 mm. Redrawn from McKay & Friedmann (1985).

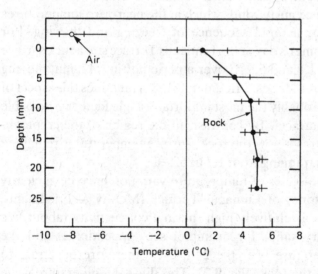

Fig. 9.3. Short-term oscillations in surface temperature on sandstone rock, Linnaeus Terrace. The temperature oscillations in the lichen zone (4.3 mm) are considerably damped and do not cross the freezing point. The bars are ranges over 40 minutes. Redrawn from McKay & Friedmann (1985).

variation was about 5 °C. Maximum temperatures were measured in the rock about three hours after maximum insolation upon the rock face.

Superimposed upon this diurnal cycle are short-term, high frequency (few minutes) oscillations caused by the cooling effect of wind gusts (Fig. 9.3). This produced rapid temperature variations at the measured rock face with an amplitude of 5–10 °C, but these oscillations were considerably damped within a few millimetres of the rock surface (Fig. 9.3).

### 9.2.5 *Nutrient supply*

Nutrient availability is unlikely to limit the population size of lithic populations in the Antarctic. Inorganic salts are released from the rock substrate and are brought in by precipitation. Nitrate, and to a less extent, ammonium often appear to be in plentiful supply in the upper few centimetres of the rock, which probably accounts for the absence of nitrogen-fixers in this habitat (Friedmann & Kibler, 1980).

### 9.3 Microbial community structure
### 9.3.1 *Phototrophic and fungal communities*
#### *Population structure*

Microalgae and fungi are closely and sometimes symbiotically associated in many of the lithic communities to the extent that they may be considered microscopic lichens. Discussion of the macroscopic (plecten-chymatous) lichen communities which occur on rock surfaces in Antarctica is outside the scope of this review and the reader is referred to Dodge (1973) and Lewis-Smith (1984).

Cyanobacteria occur under the surface of sandstone rocks in the Dry Valleys (Fig. 9.4) and were the first antarctic endoliths to be described (Friedmann & Ocampo-Friedmann, 1976). These communities consist of strains of the coccoid genus *Chroococcidiopsis*, but subsequent investigations indicate that this cyanobacterial community is not widespread.

The predominant endolith community, at least in the Dry Valleys, appears to be a symbiotic lichen association between unicellular green algae and filamentous fungi. The loose filaments and groups of unicells grow between and round the crystals of the rock. As in many lichen associations, the fungi produce appressoria, hyphal swellings that surround the algal symbiont, or haustoria, specialised hyphae that penetrate the algal cell. These structures presumably allow an exchange of materials between the two symbionts. Certain chemical compounds that are characteristic of lichen associations – norstictic acid and gyrophoric acid – have

(a)

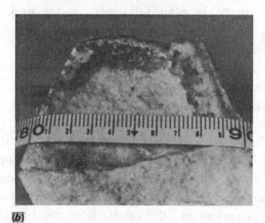

(b)

Fig. 9.4. Two lithic communities in the Dry Valleys. (*a*) Cryptoendolithic layer of cyanobacteria in sandstone. The blue-green coloured layer is 1.5 mm thick and 5 mm below the smooth rock surface. (*b*) Chasmolithic community in marble. A 10 mm wide black lichen layer overlies a 5 mm, blue-green layer of cyanobacteria. In both photographs the rocks have been broken open to expose the microbial communities.

been identified in these unusual endolithic communities. The cryptoendoliths do not generally possess sexual reproductive structures, but these may be produced when they become exposed to cracks on the rock surface and experience a favourable microenvironment. From these sexual features five lichen genera of endoliths have been identified: *Buellia, Lecidea, Carbonea, Lecanora* and *Acarospora* (Hale, 1987). The algal symbionts include at least two genera – *Trebouxia*, which is a common constituent of lichens at all latitudes, and *Pseudotrebouxia* (classified as *Trebouxia* by some phycologists, I. Friedmann, pers. comm.). A separate free-living

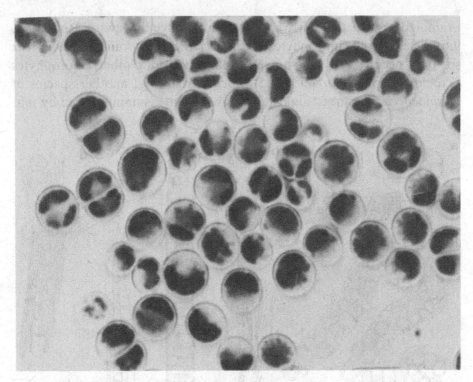

Fig. 9.5. The free-living green alga *Hemichloris antarctica* which forms a bright green layer under cryptoendolithic lichens in the Dry Valleys. From Tschermak-Woess & Friedmann (1984) by permission of the authors and Phycologia.

green alga, *Hemichloris antarctica*, often forms a bright green layer beneath the lichen (Fig. 9.5), and is frequently accompanied by *Chroococcidiopsis* or *Gloeocapsa* (Tschermak-Woess & Friedmann, 1984).

These distinctive non-foliose lichens may also colonise the rock fissure habitat (Fig. 9.4), or may become chasmolithic as the rock surface begins to decompose partly due to the activity of the micro-organisms themselves. Chasmolithic lichens of the general *Buellia* and *Lecidea* have been identified in the Dry Valleys. These species are abundant in the layers of exfoliating rock, as well as vertical fissures.

Many of the chasmoliths occur without associated fungi. For example, at Marble Point on the coastal edge of the southern Victoria Land Dry Valleys outcrops of weathered Koettlitz marble are generally traversed by numerous fractures. These are heavily colonised by cyanobacteria, microscopic chlorophytes, and *Heterococcus*, the xanthophyte typical of many soils throughout the world (Friedmann 1982).

Eight species of microalgae dominated the chasmolithic habitats examined by Broady (1981b) at Vestfold Hills and Mawson Rock (Fig. 9.6). The coccoid chlorophytes *Prasiococcus calcarius* and *Desmococcus* sp. only occurred in salt-sprayed areas while four other chlorophytes (*Trochiscia* sp., *Chlorella* sp., *Stichococcus* sp. and another species of *Desmococcus*) occurred almost exclusively in areas uninfluenced by salt.

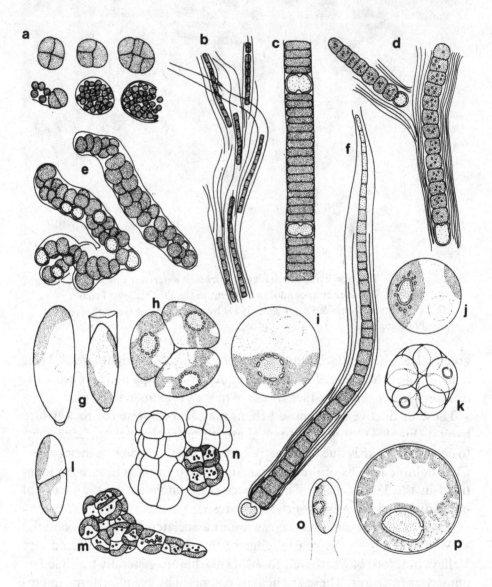

Fig. 9.6. Chasmolithic species from the Vestfold Hills. (*a–f*) cyanobacteria, (*g–p*) chlorophytes. Reproduced from Broady (1981b) by permission of the author and Academic Press Australia.

Large areas of the Vestfold Hills are affected by seaspray, and this algal group, especially *Prasiococcus* (Fig. 8.5), was frequently recorded in sample material from there. Two cyanobacteria occurred under both conditions – the coccoid species *Chroococcidiopsis* (Fig. 9.6a), and flexuous unbranched and occasionally false-branched filaments of the genus *Plectonema*. The relative abundance of each species appeared to be also related to moisture availability. *Chroococcidiopsis* sp. was common in sites that received meltwater as a thin film over the rock for at least part of summer, while *Desmococcus* (Fig. 9.6*m,n*) was often found over large areas of rock that did not experience meltwater flows.

The chasmoliths at Mawson Rock and the Vestfold Hills generally comprised a single stratum of predominantly chlorophytes or cyanophytes, but occasionally a distinct zonation was observed. These vertically structured communities consisted of an inner blue-green zone of *Chroococcidiopsis* and an outer green layer of predominantly *Prasiococcus* in the vertical cracks or *Desmococcus* under the flakes.

The vertical zonation of microbial constituents is much more pronounced in the *Hemichloris*-containing endolithic communities of the Dry Valleys region. The upper few millimetres of rock are free of microorganisms. An upper black stratum well below the rock surface consists of darkly pigmented masses of fungal hyphae enclosing groups of algal cells (*Trebouxia*). This stratum overlies a white zone of fungal mycelium where the population of algal symbionts is sparse. Beneath this region is a green layer of the free-living chlorophyte *Hemichloris antarctica*. There are fungal hyphae in this bottom stratum, but they do not appear to form a lichen association with the alga. An additional layer of cyanobacteria (*Chroococcidiopsis, Gloeocapsa*) may underlie the green algal stratum.

Microalgae and associated organisms living on the exposed rock-face experience the hostile extremes of cold and desiccation. The species diversity of this epilithic microbiota is predictably limited. If present, the epilithic community forms a thin dark-brown or black crust which is almost entirely composed of cyanobacteria (Broady, 1981c, 1986). The important genera are *Gloeocapsa*, *Calothrix* and thin-trichome members of the Oscillatoriaceae. Species similar to *Myxosarcina* and *Chroococcidiopsis* have been observed, but these are probably cells derived from *Gloeocapsa* after desiccation of its mucilaginous sheath. This sheath is characteristic of all of the epilithic cyanobacteria and is often darkly pigmented orange, violet or brown. It probably performs a variety of important functions including providing shade from bright light and protection against rapid desiccation.

The epilithic crusts require a small film of meltwater at some stage during summer and are therefore often found downslope of snowdrifts (Broady, 1986). However, the cyanobacterial species are intolerant of high salinity and are absent from rock surfaces experiencing seaspray or wind-drift from saline lakes.

Sublithic algal communities develop on the undersurface of translucent rocks such as quartz and marble. Of all the rock habitats, the sublithic environment is probably the most favourable for microbial growth. This is reflected in a higher species diversity, although mosses and lichens are rare (Broady, 1981d, 1986). The dominant species are cyanobacteria, especially Oscillatoriaceae, but *Desmococcus* and *Prasiococcus* are frequently recorded, as well as a range of associated species in lesser abundance.

## Biomass structure

Lithic microbial populations are relatively common in certain areas. In the Vestfold Hills area chasmoliths occurred in 46 (75%) out of 61 1-km$^2$ grid

Table 9.1. *Typical biomass levels of phototrophs (as measured by chlorophyll* a) *in the lithic and soil communities of Antarctica*

| Community | Biomass ($\mu$g chl $a$ cm$^{-2}$) |
|---|---|
| *Cryptoendoliths in southern Victoria Land* | |
| cyanobacteria (in sandstone) | 2–6 |
| lichen (in sandstone) | 0.1–3 |
| *Adjacent soil* | 0 |
| *Chasmoliths in southern Victoria Land (in marble)* | 10 |
| *Subliths in Vestfold Hills*[a] | |
| dry soil beneath | 12 ± 6 |
| moist soil beneath | 6 ± 4 |
| *Adjacent soil* | 0–0.4 |
| *Algal felts, Vestfold Hills*[a] | |
| Nostoc | 54 ± 26 |
| Oscillatoriaceae | 203 ± 70 |
| *Prasiola crispa* | 533 ± 152 |
| *Moss, Vestfold Hills*[a] | 77 ± 20 |

[a] Vestfold Hills data are means of 8 samples ± standard deviation.
Derived from Friedmann *et al.* (1980) and Broady (1981d).

sites, and within each area occupied 2–21% of the total rock surface area. This cover was much greater than either epilithic algae or lichens (Broady, 1981b).

Endolithic communities at the Linnaeus Terrace in the Dry Valleys had a standing biomass of 32–177 g m$^{-2}$ of rock surface, based on Kjeldahl nitrogen, ATP and chlorophyll *a* determinations. These biomass levels are comparable with temperate desert endolithic communities – for example from 38 to 185 g m$^{-2}$ in the Negev and Sonoran deserts (Friedmann & Ocampo-Friedmann, 1984).

The concentration of chlorophyll *a* in sublithic and cryptoendolithic communities is relatively low by comparison with surface terrestrial populations in Antarctica (Table 9.1) particularly relative to damp soil communities (see Chapter 8). But in some areas (e.g. alpine areas in southern Victoria Land, and the Vestfold Hills) the wide distribution of these lithic communities, where conditions are unfavourable for other biota, indicates that they may be of considerable importance.

### 9.3.2  *Bacterial communities*

Colourless bacteria are commonly found in the endolithic habitat where they presumably are heterotrophic and operate as decomposers. Unidentified bacteria have been recorded in association with the cyanobacterial endoliths, and regularly accompany the endolithic lichens as microscopic colonies of rod-shaped cells (Friedmann, 1982).

### 9.4  **Microbial processes**
### 9.4.1  *Phototrophic production*

$CO_2$-exchange assays have provided the first estimates of photosynthesis and respiration by the cryptoendolithic communities in the Dry Valleys (Kappen & Friedmann, 1983). Net photosynthesis was measured down to the minimum experimental temperature of $-4.2\,°C$, but rose with increasing temperature to an optimum in the range 2–6 °C (Fig. 9.7). This temperature optimum rose slightly with increasing photon flux density from 217 to 551 $\mu E$ m$^{-2}$ s$^{-1}$. These temperature responses are comparable with those of fruticose and foliose lichens of cold climates and do not suggest any unusual adaptation to the extreme cold of the Dry Valley alpine desert environment. For this cryptoendolithic community the maximum photosynthetic rate was generally between 0.05 and 0.1 mg $CO_2$ (mg chl*a*)$^{-1}$ h$^{-1}$. The ratio of chlorophyll *a* : *b* lay in the range 1.9–3.0 which is similar to values for other lichens.

A predicted temperature compensation point (temperature at which

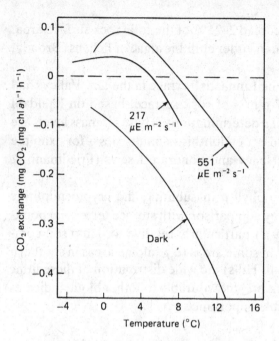

Fig. 9.7. Influence of temperature on cryptoendolith photosynthesis and respiration. The community sampled from Linnaeus Terrace was incubated in the dark, and at 217 and 551 $\mu E$ m$^{-2}$ s$^{-1}$. Redrawn from Kappen & Friedmann (1983).

photosynthesis balances respiration) for these endolithic lichens of $-6\,°C$ would mean that net photosynthesis could continue at positive rates for 13 h or less of a typical mid-summer day in the Dry Valleys. A temperature optimum of $8\,°C$ was predicted for full sunshine ($1420\ \mu E$ m$^{-2}$ s$^{-1}$) conditions. This temperature would be rarely met in the natural environment, and Friedmann *et al.* (1987) consider low temperatures to be the most important single factor restricting the biological activity of these endolithic communities.

Cold climate lichens often demonstrate high respiration rates at warm temperatures, which results in a very low ratio of maximum $CO_2$ uptake: dark $CO_2$ production. This quotient was extremely low for the endoliths, lying between 0.1 and 0.2 at $10\,°C$. One of the samples demonstrated no net photosynthesis above $8\,°C$ and the two other samples had an upper compensation point at about $14\,°C$ under high photon flux densities. This sensitivity to high temperature is even greater than for other antarctic lichens; three lichens from Cape Hallett, for example, had quotients between 0.08 and 0.19 at $20\,°C$. For a clear, warm, calm day Kappen *et al.*

(1981) estimate that photosynthesis would operate for a total of 13.5 h, with 6.5 h at rates not limited by temperature.

The measured photosynthetic rates per unit chlorophyll *a* for this endolithic community were at least an order of magnitude lower than epilithic antarctic lichens which are adapted to much brighter light intensities. In the sealed endolithic environment net photosynthetic rates may be even further reduced by the slow rate of $CO_2$ diffusion through the rock surface crust. $CO_2$ recycling by the heterotrophic fungal symbiont may be an important feature regulating photosynthesis in the endolithic ecosystem. Shelf-shading may further constrain the space available for microalgal growth.

### 9.4.2 Biogenic exfoliation

The endolithic community may accelerate the chemical weathering process within certain rocks, particularly sandstones. It appears that although endolithic lichens enter the rock surface by growing through pore spaces, once established they produce substances that solubilise the material cementing the sandstone grains together. The upper crust of rock then peels away ('exfoliates') and the hyphae penetrate deeper to produce a new lichen zone. During this process the lichens mobilise iron compounds which are deposited both below the lichen zone and at the surface of the rock. This biogenic weathering results in a pattern of irregular surface crusts strained with iron and patches of white and dark brown where the rock has been recently exposed (Friedmann, 1982).

### 9.5 Trophic structure

Rock ecosystems have an extremely simple trophic structure, although the antarctic endolithic microbiota appear to be more diverse than in hot temperate deserts. Eukaryotic and prokaryotic algal cells are the only phototrophs and this production supports heterotrophic bacteria and fungi. Algal and fungal cells are physically connected in the lichen associations, but even in the non-lichen assemblages there may be a close functional coupling between each component; for example, $CO_2$ regenerated by bacteria and fungi may be the dominant carbon source for algal photosynthesis within the endolithic pore spaces. There are no herbivores or predators in the endolithic environment although presumably mites and collembolans may be occasionally active on the outer rock surfaces. The endolithic habitat is partially sealed and well-insulated from the influence of the outside flora, fauna and environment and there is some evidence that an endemic microflora may have evolved under these unique conditions. *Hemichloris* has not been recorded elsewhere in the world, and a new yeast has been isolated from the endolith communities (Vishniac, 1985). These organisms may have been important life forms in the *nunataks* (ice-free mountain tops projecting through the ice mass) which remained exposed during glaciations, perhaps representing 'the last footholds of life in a gradually deteriorating environment' (Friedmann & Ocampo-Friedmann, 1984). As such the endolithic ecosystem has attracted considerable interest as a model for the detection of life forms on other planets.

# 10

# Microbial strategies in Antarctica

## 10.1    Introduction

Extremes of low temperature, salt concentration and aridity have
excluded all but microbial life forms from many antarctic environments. In
the absence of severe grazing pressure from animals and without strong
competition from a wide range of other microbial species, certain micro-
organisms grow to a prolific biomass in parts of the region. To live within
Antarctica, however, they require an unusual set of tolerances, or adaptive
strategies that minimise the impact of the environmental extremes. These
cellular and life cycle strategies can also be found at temperate latitudes,
but the physical and chemical conditions which are imposed in Antarctica
have selected for a particularly robust and resilient microbial flora. This
chapter examines microbial responses to characteristics of the antarctic
environment including freezing temperatures, high salinity, continuous
winter darkness, and prolonged high or low radiation. The cold temperatures of Antarctica have a wide-ranging influence in all of the microbial
ecosystems of this region, and the responses to this environmental feature
are therefore examined in greatest detail.

## 10.2    Low temperatures

Antarctica contains some of the coldest environments to be found
on Earth, and low temperatures must exert a fundamental control on
microbial life processes throughout the region. Cold influences biological
activity in various ways: through changes in biochemical reaction rates,
shifts in the stability of some cellular components and through the disruptive effects of freezing. This section describes how these low temperature effects operate on cells in general, the temperature responsive-
ness of antarctic microbiota, and the potential strategies of adaptation in
Antarctica.

### 10.2.1 Cold avoidance

Certain microbial communities in Antarctica escape the extreme cold of this climatic zone by their choice of habitat. A few antarctic environments remain continuously warm throughout the year: fumeroles and geothermally active sites; deep strata in solar-heated lakes; homeothermic animals including man; and the ever increasing number of human settlements. All of these habitat types are rare, and with the exception of the human environments which experience frequent exchanges with the rest of the world, microbial colonisation must proceed very slowly.

It is testimony to the enormous range of tolerances and flexibility of certain microbial groups that some taxa are represented in both the hot and cold extremes of the antarctic environment. Members of the cyanobacterial family Oscillatoriaceae, for example, are found in antarctic streams below 0 °C, the deep waters of Lake Vanda (18 °C) and in the heated soils on Mt Melbourne and Mt Erebus (Ross Sea sector) at temperatures of up to 60 °C.

The microbial communities of ice-covered lakes inhabit cold waters but generally experience a warmer and less variable thermal environment than in the surrounding catchment. Water temperatures in the lakes of the Dry Valleys (Appendix 1) persist throughout winter at temperatures 40–80 °C warmer than the overlying air. On Signy Island, lakewater temperatures beneath the ice in winter lie near 0 °C, but air temperatures drop to below −15 °C (Appendix 1). The dozens of meltwater streams in the vicinity of McMurdo Sound often ice up for several hours over the diel cycle, but freezing conditions within the microbial mats beneath this insulating ice cover probably occur infrequently throughout the period of summer streamflow. Other antarctic environments may also buffer their microbial populations from large shifts in temperature over the 24 h cycle, or at shorter time intervals. The heat stored within rock and soil, for example, will dampen temperature fluctuations experienced by the microbiota of these habitats (Fig. 10.1) although the storage capacity per degree Celsius (heat capacity) is much less for these materials than for water (Appendix 2). The marine plankton and sediment micro-organisms inhabit extremely stable environments which, although cold, rarely drop below −2 °C. By contrast the snow microbiota and upper sea-ice communities experience rapid changes in temperature, and may possibly have to endure frequent cycles of freezing and thawing.

### 10.2.2 Cellular responses to chilling

The primary response of micro-organisms to low temperature is a slowing of all physiological processes. This temperature effect is generally

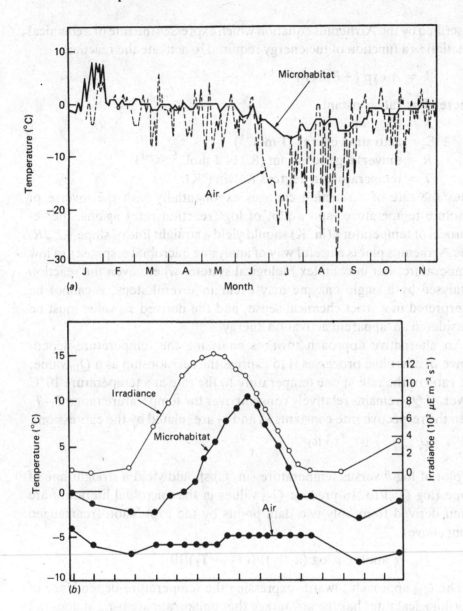

Fig 10.1. Microclimatic influences on temperature. (*a*) the annual variation in soil temperature on a rocky outcrop between the sea and the ice cap on Anvers Island. (*b*) The diurnal variation in soil temperature beneath a quartz stone in the Vestfold Hills. In each microhabitat the temperature lies well above that of the surrounding air and freezing conditions are much less frequently experienced. Redrawn from Baust & Lee (1980) and Broady (1981d).

described by the Arrhenius equation which expresses the rate of a chemical reaction as a function of the energy required to activate the reactants:

$$k = A \exp(-E_a/RT)$$

where $k$ = rate constant

$A$ = constant

$E_a$ = activation energy $(J\ mol^{-1})$

$R$ = universal gas constant $(8.314\ J\ mol^{-1}\ {}^{\circ}K^{-1})$

$T$ = temperature in degrees Kelvin $({}^{\circ}K)$.

Thus the rate of a reaction changes exponentially with the inverse of absolute temperature, and a plot of log (reaction rate) against the reciprocal of temperature (in ${}^{\circ}K$) should yield a straight line of slope $-E_a/R$. The Arrhenius plot is a useful way of analysing microbial responses to low temperature, but in complex biological systems where even the reaction catalysed by a single enzyme may occur in several steps, it cannot be interpreted in a strict chemical sense, and the derived $E_a$ value must be considered an 'apparent activation energy'.

An alternative approach towards analysing the temperature-dependence of microbial processes is to express the relationship as a $Q_{10}$ value: the ratio of the rate at one temperature to the rate at a temperature $10\,{}^{\circ}C$ lower. If $Q_{10}$ remains relatively constant over the temperature range $T_1-T_2$ then the respective rate constants $k_1$ and $k_2$ are related by the expression:

$$k_2 = k_1\ Q_{10}^{((T_2-T_1)/10)}$$

A plot of log $k$ versus temperature (in ${}^{\circ}C$) should yield a straight line of slope (log $Q_{10}$)/10. In practice $Q_{10}$ values in the microbial literature are often derived from only two data points by the expression (rearranged from above):

$$Q_{10} = \text{antilog}\ [(\log\ (k_2/k_1))/((T_2 - T_1)/10)]$$

The $Q_{10}$ approach towards expressing the temperature-dependence of physiological rates has the advantage that unlike 'apparent $E_a$' it does not imply any mechanistic explanation. The two parameters are fundamentally different, but one may be derived from the other by the relationship:

$$-E_a = (RT(T + 10)\ \ln\ Q_{10})/10$$

For the temperature range 0–20 °C the relationship may be approximated to within 10% by:

$$E_a = -70\ \ln\ Q_{10}$$

where $E_a$ is in KJ $mol^{-1}$.

Arrhenius plots for biochemical and physiological processes often demonstrate a 'break' or abrupt shift in slope. These changes in gradient imply a change in the rate-limiting step in a series of reactions, or a shift in the catalytic activity of the system. They may be observed even for pure enzymes – for example the $Q_{10}$ for the carboxylation enzyme, ribulose 1,5 bisphosphate carboxylase (RuBPCase), changed from 2.2 to 3.3 at temperatures below 15 °C (Badger & Collatz, 1977) suggesting that low temperatures induced a conformational change in the enzyme proteins which modified its catalytic properties. Biophysical shifts in membrane structure may be an explanation of the change in slope of the Arrhenius plot for some biological systems, and appear to be an important response to chilling.

Cell membranes consist of a fluid bilayer of lipid in which are embedded the proteins responsible for transport, phosphorylation, light harvesting and other membrane functions. The proteins may be partially submerged in the lipid or may completely span the bilayer, and are free to diffuse in the plane of the membrane. The fluidity of the bilayer appears to be an important requirement for the activity of membrane proteins, but at low temperatures the bilayer solidifies and many membrane functions are impaired. The bilayer is composed of a complex mixture of different types of lipid, and the phase transition from a 'liquid–crystalline' state to a gel occurs over a range of temperatures. Within this range the two different phases may co-exist in different domains of the membrane, and proteins may be concentrated in the fluid phase or expelled from the membrane. Even above this range, changes in membrane fluidity may closely regulate the activity of certain proteins – the apparent $E_a$ of succinate oxidase in mitochondrial membranes, for example, increased with increasing bilayer fluidity (McMurchie & Raison, 1979). The temperature at which a lipid mixture solidifies is very much dependent upon its fatty acid composition. The degree of saturation of the fatty acids dramatically affects fluidity – saturated acids are less fluid than unsaturated acids. Small changes in fatty acid composition can have a large effect on the phase transition temperature and so bulk compositional analysis of membranes for their level of unsaturation may not be very informative.

Low temperature effects on the physical properties of membranes is currently regarded as the most attractive explanation for impaired biological activity in response to chilling. However, various other types of 'chilling injury' may operate at low temperature and effectively exclude certain micro-organisms from Antarctica. Various enzymes, for example, are unstable at low temperature. The proteins spontaneously unfold and multi-subunit structures disassemble into inactive components which do not reassociate when the system is rewarmed (e.g. phosphofructokinase; Dixon, Franks & Rees, 1981). Microtubules which form the structural units of flagella and the cytoskeletal system disassemble at low temperatures. In most plant and animal cells they depolymerise at about 4 °C. The biological activity of proteins is very much dependent upon the extent of hydrogen bonding with $H_2O$ molecules, and this surrounding water structure can be perturbed by extreme low as well as high temperatures (Franks, 1985). Changes in membrane permeability associated with phase transitions may expose the cell to toxic compounds, or allow the accumulation of certain electrolytes to inhibitory levels. For example, many cellular processes are extremely sensitive to free carboxylic acids (see Steponkus, 1981).

At high temperatures changes in membrane stability and protein denaturation slow microbial processes and eventually lead to cellular death. This combination of temperature effects produces abrupt changes in the slope at several points on the Arrhenius curve (Fig. 10.2). Over the range $T_1$–$T_2$ the physiological process conforms to the Arrhenius relationship. Below the break point at $T_1$ membrane transitions and/or other effects of chilling impair metabolism; above $T_2$ (sometimes called the physiological optimum) high temperature begins to injure cellular components, but reaction rates continue to respond to warming. Above $T_3$ (called the 'growth optimum' for Arrhenius plots of specific growth rate) the injurious effects of high temperature dominate the response curve and physiological rates begin to fall.

There are certain biologically important processes that are much less responsive to temperature over the ambient range than cellular biochemistry. Photochemical reactions that are limited by the photon flux density ($I$) are largely temperature-independent. Thus temperature has little influence on photosynthesis ($P$) over the light-limited portion ($a$) of the $P$ versus $I$ curve by comparison with its effect at light saturation ($P_{max}$) where the rate-limiting steps are biochemical rather than photochemical. Diffusive processes are also less sensitive to ambient temperature than biochemical reactions. The relative change in rates of molecular diffusion ($D_1$, $D_2$) between two temperatures ($T_1$, $T_2$, in degrees Kelvin) is given by:

$$D_1/D_2 = T_1\eta_2/T_2\eta_1$$

where $\eta_1$ and $\eta_2$ are the viscosities of the surrounding medium at the two temperatures. Over the temperature range 25 to −5 °C, for example, the viscosity of water increases from 0.89 to 2.14 centipoise. From the above relationship, diffusion rates at −5 °C would decrease to 37% of the rates at

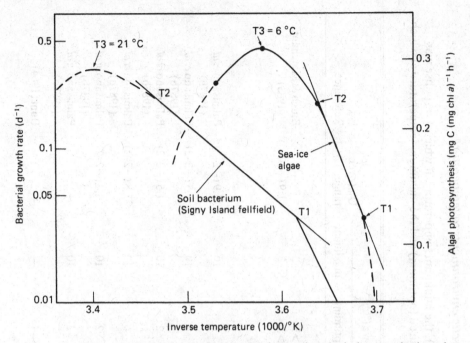

Fig. 10.2. A hypothetical Arrhenius plot for sea-ice algal photosynthesis and a portion of the Arrhenius curve for growth of a soil bacterium. The sea-ice data points are for the congelation-ice algal community from McMurdo Sound (Palmisano *et al.* 1987b) and the bacterial data are for *Corynebacterium* from Signy Island (Ellis-Evans & Wynn-Williams, 1985). $T_1$, $T_2$ and $T_3$ are described in the text.

25 °C. By comparison, biochemical reactions with a $Q_{10}$ of 2.0 would fall to 12.5% of the rates at 25 °C over this temperature range. This much greater influence of chilling on reaction rates than diffusion may offset the effect of the long diffusion pathway for nutrients passing through the protective mucilages that encapsulate certain microbial cells in Antarctica (see Section 10.2.6). It may also confer an advantage to micro-organisms at low temperatures over higher plants and animals which are more dependent upon active metabolism for their nutrition.

### 10.2.3 *Antarctic temperature response curves*

Photosynthesis, bacterial growth rates and other microbial processes in most antarctic ecosystems appear to be depressed by the low environmental temperatures, with optima generally at least several degrees above the ambient range (Table 10.1). $Q_{10}$ values for these processes have often exceeded 2.0 (Table 10.1) indicating a strong response to small increases in temperature.

Table 10.1. *Response of microbial populations from or in antarctic environments to temperature*

Temperature data include the natural habitat temperature (ambient), the minimum temperature at which a response was measured, the temperature at which the greatest response was measured (optimum) and the maximum temperature at which a response was measured.

| Community/population | Response measured | Temperature (°C) | | | | $Q_{10}$ (over range, °C) | Reference |
|---|---|---|---|---|---|---|---|
| | | ambient | minimum | optimum | maximum | | |
| *Freshwater ice* | | | | | | | |
| *Chlamydomonas* sp. | motility | <0 | −14 | >3 | 3 | — | Burch & Marchant (1983) |
| *Pyramimonas gelidicola* | motility | <0 | −10 | >3 | 3 | 1.9 (−10–0) | |
| (Both from lake ice cores, Vestfold Hills) | | | | | | | |
| *Sea ice* | | | | | | | |
| Congelation ice algae (McMurdo Sound) | $P_{max}$ | <−1.8 | −2 | 6 | 10 | 5.9 (−2–2) | Palmisano *et al.* (1987b) |
| | α | <−1.8 | −2 | 2 | 10 | 3.5 (−2–2) | Palmisano *et al.* (1987b) |
| | β | <−1.8 | −2 | 2 | 10 | 1.5 (−2–2) | Palmisano *et al.* (1987b) |
| Platelet ice algae (McMurdo Sound) | $P_{max}$ | <−1.8 | −2 | 6 | 10 | 27.8 (−2–2) | Palmisano *et al.* (1987b) |
| | α | <−1.8 | −2 | 2 | 10 | 3.4 (−2–2) | Palmisano *et al.* (1987b) |
| | β | <−1.8 | 2[a] | 2 | 10 | — | Palmisano *et al.* (1987b) |
| Platelet ice algae (McMurdo Sound) | $P_{max}$ | <−1.8 | — | 10–15 | — | — | Bunt (1964) |

| | | | | | | |
|---|---|---|---|---|---|---|
| *Southern Ocean* | | | | | | |
| *Chaetoceros* sp. (culture) | growth | 0–5 | 3 | 7–10 | 10[b] | 8.6(3–5) | Jacques (1983) |
| *Nitzschia turgiloides* (culture) | growth | 0–5 | 3 | 5 | 5[b] | 26.1(3–5) | Jacques (1983) |
| *Fragilariopsis kerguelensis* (culture) | growth | 0–5 | 3 | 3 | 7[b] | — | Jacques (1983) |
| Phytoplankton, Scotia Sea | $P_{max}$ | –0.8–1.0 | — | 7 | 28 | 2.7(0–5) | Neori & Holm-Hansen (1982) |
| Phytoplankton, Scotia Sea | $P_{max}$ | –0.8–1.0 | –1.5 | >8 | 8 | 4.6(–1.5–0) | Tilzer *et al.* (1986) |
| | | –0.8–1.0 | –1.5 | 5 | 8 | 2.6(–1.5–5) | |
| Phytoplankton, Drake Passage | dark respiration | –0.8–1.0 | –2 | 7–8 | >8 | 2.3(–2–8) | Tilzer & Dubinsky (1987) |
| | | –0.8–1.0 | –2 | >8 | 8 | 11.9(–2–8) | Tilzer & Dubinsky (1987) |
| Bacterioplankton, 2 stations | glutamate uptake | –0.8–1.0 | –3 | >16 | 16 | 6.0(–3–6) | Morita, Griffiths & Hayasata (1977) |
| | | –0.8–1.0 | –3 | 11 | 16 | 2.6(–3–6) | Morita *et al.* (1977) |
| Picoplankton, Scotia Sea | Growth rate by ATP increase | 5 | 0 | >15 | <15 | 3.5(0–10) | Hanson & Lowery (1985) |
| Bacterioplankton, Scotia Sea | growth | –1–1 | –1 | 7 | — | 1.6(–1–7) | Azam, Ammerman & Cooper (1981) |
| *Marine benthos* | | | | | | |
| *Notodendrodes antarcticos* (foraminifer from Ross Sea sediments) | uptake of organics | –1.8 | –1.5 | >5–10 | 10 | 1.5(–1.5–5) | De Laca (1982a, b) |

*continued*

Table 10.1 *contd.*

| Community/population | Response measured | Temperature (°C) | | | | $Q_{10}$ (over range, °C) | Reference |
|---|---|---|---|---|---|---|---|
| | | ambient | minimum | optimum | maximum | | |
| *Lakes, pools and streams* | | | | | | | |
| *Chlamydomonas globosa* (culture) | growth | 0–5 | 5[a] | 18 | 20 | 1.8 (5–10) | Seaburg *et al.* (1981) |
| *Chlamydomonas intermedia* (culture) | growth | 0–5 | −1 | 18 | 18[b] | 3.4 (−1–5) | Seaburg *et al.* (1981) |
| *Chlamydomonas subcaudata* (culture) | growth | 0–5 | −1 | 12.5 | 18[b] | 3.9 (−1–5) | Seaburg *et al.* (1981) |
| *Chloromonas alpina* (culture) | growth | 0–5 | −1 | 12.5 | 18[b] | 11.2 (−1–5) | Seaburg *et al.* (1981) |
| Phytoplankton, Skua Pond | photosynthesis under bright light | 8 | 4 | >14 | 14 | 4.9 (4–14) | Goldman *et al.* (1963) |
| Phytoplankton, Algal Pond | | 6 | 4 | >14 | 14 | 8.9 (4–14) | Goldman *et al.* (1963) |
| Phytoplankton, Skua Pond | photosynthesis in dim light | 8 | 2.5 | >26 | 26 | 2.2 (2.5–26) | Goldman *et al.* (1963) |
| Phytoplankton, Algal Pond | | 6 | 2.5 | >26 | 26 | 1.4 (2.5–26) | Goldman *et al.* (1963) |
| Benthic mats, Skua Pond | $P_{max}$ | >8 | 4 | >14 | 14 | 2.3 (4–14) | Goldman *et al.* (1963) |
| Benthic mats, Algal Pond (both dominated by Oscillatoriaceae) | $P_{max}$ | >6 | 4 | >14 | 14 | 1.7 (4–14) | Goldman *et al.* (1963) |
| *Halomonas subglaciescola* (culture of halotolerant bacterium from Vestfold Hills) | growth | −7–1 | −5 | — | 25[b] | — | Franzmann, Burton & McMeekin (1987a) |
| Lake bacterial isolates from Signy Island | growth | 0–5 | | 14–19 | 21–24[b] | — | Ellis-Evans & Wynn-Williams (1985) |

| | | Env. temp. | min. | opt. | max. | Reference |
|---|---|---|---|---|---|---|
| *Phormidium* mat (Fryxell Stream) | $^{14}C$ photosynthesis | 0->5 | 0 | >15-25 | 25 | Vincent & Howard-Williams (1987) |
| | respiration | 0->5 | 0 | >10 | 10 | Vincent & Howard-Williams (1988) |
| | glucose catabolism | 0->5 | 0 | >10 | 10 | Vincent & Howard-Williams (1988) |
| *Nostoc* mat (Fryxell Stream) | $^{14}C$ photosynthesis | 0->5 | 0 | >10 | 10 | Vincent & Howard-Williams (1988) |
| | respiration | 0->5 | 0 | >10 | 10 | Vincent & Howard-Williams (1988) |
| | glucose catabolism | 0->5 | 0 | >10 | 10 | Vincent & Howard-Williams (1988) |
| *Nostoc commune* (4 isolates) | growth | 0->5 | 2 | | 30[b] | Seaburg *et al.* (1981) |
| *Nostoc commune* (2 isolates) | growth | 0->5 | 5[a] | | 25[b] | Seaburg *et al.* (1981) |
| 5 other isolates of this species had slightly different minima and maxima | | | | | | |
| **Soil** | | | | | | |
| *Planococcus* isolate A4a (halotolerant bacterium from Dry Valleys) | growth | -50-20 | | 25 | 40 | Miller *et al.* (1983) |
| Soil bacterial isolates from Signy Island | growth | — | | 15-29 | 23-33[b] | Ellis-Evans & Wynn-Williams (1985) |
| *Nostoc commune* | photosynthesis | -50-20 | -5[a] | >5 | 5 | Becker (1982) |
| *Prasiola crispa* | photosynthesis | -50-20 | -15[a] | (-15-5) | 5 | Becker (1982) |
| **Rock** | | | | | | |
| Cryptoendolithic lichens | net photosynthesis | -12-7 | -4 | 4-8 | 8-15[b] | Kappen & Friedmann (1983) |
| | respiration | -12-7 | 0 | >12 | 12 | Kappen & Friedmann (1983) |
| Cryptoendoliths | $CO_2$ fixation into lipid | -12-7 | 0 | ≥15 | 15 | Vestal (1985) |

[a] indicates that tests were also conducted below this temperature.

[b] indicates that tests were also performed above this temperature.

True psychrophily, that is genetic adaptation to growth at low temperatures, seems to be relatively uncommon in the terrestrial and freshwater antarctic environments. Even in the stable cold temperatures of the Southern Ocean many of the bacteria seem to be psychrotrophic (i.e. capable of growth at low temperatures, but with growth optima at much warmer temperatures) rather than psychrophilic, although the extent of psychrophily seems higher than in the Arctic Ocean (Atlas & Morita, 1987). Some studies have shown that many bacterial isolates and some phytoplankton from the Southern Ocean cannot grow above 10 °C (see Section 5.4.2). Even these species, however, are highly responsive to warming several degrees above the ambient range and show no evidence of very rapid growth at low temperature. Bacterial isolate AF-1 from Heywood Lake (Signy Island) had a temperature optimum of 10 °C, a maximum limit of <20 °C, but at ambient temperatures of 0 °C had a long generation time of 23 h (Herbert & Bell, 1977). Temperature adaptation appears to be manifested as an impaired ability to grow at high temperature, and a tolerance of the sub-optimal, ambient temperature regime.

The temperature optima for enzyme activity in biological systems typically lie well above the ambient range (Franks, 1985). This appears to be true for some but not all microbial enzyme systems in Antarctica. For example, Reichardt & Dieckmann (1985) examined the temperature response curves for enzymes produced by heterotrophic bacteria during the decomposition of antarctic marine macroalgae and found that amylase and protease had optima around 25 °C while cellulase activity was maximal at 0 °C (Fig. 10.3).

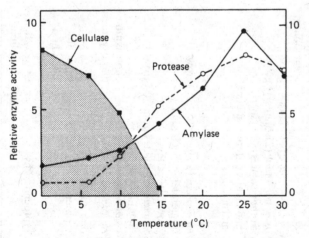

Fig. 10.3. Temperature characteristics of bacterial enzymes responsible for the decomposition of the macroalga *Himantothallus grandifolius* from Admiralty Bay, King George Island. Redrawn from Reichardt & Dieckmann (1985).

Simple physiological responses may be highly misleading in interpreting the community response to changes in temperature, especially in mixed assemblages (*consortia*) of heterotrophs and phototrophs. For example, respiration rates in the cryptoendolithic lichen communities within Dry Valley rocks were highly responsive to temperature. Above 8 °C, however, there was no further stimulation of photosynthesis while respiration continued to increase, and the community then experienced a net carbon loss (Kappen & Friedmann, 1983). Similarly low temperature compensation points (where net photosynthesis equals zero) have been observed in sea-ice communities (Bunt, 1968) and stream cyanobacterial mat communities (Vincent & Howard-Williams, 1988).

### 10.2.4 *Adaptation to chilling*

Three general strategies of biochemical adaptation may allow micro-organisms to cope with the stresses imposed by chilling (Hochachka & Somero, 1984):

1. Qualitative changes in cellular constituents such as new proteins or new lipid mixtures that perform well at low temperature.
2. Increases in the cellular concentration of enzymes and other constituents to compensate for the temperature depression of the corresponding reaction.
3. Intermediate responses of the metabolic system to control or minimise the injuries imposed by chilling.

Cold-adapted microbial populations in the Antarctic probably have membrane lipid mixtures that undergo phase transition at very low temperature. Cold injury in the cyanobacterium *Anacystis nidulans* has been attributed to phase separations in the plasmalemma (Ono & Murata, 1982), and thus cyanobacteria as a group cannot be considered generally pre-adapted to cold environments. Even related microbial species in the same genus may differ substantially in their ability to survive chilling. For example, six strains of *Amoeba* sp. from temperate latitudes, including three members of the same species, varied in their recovery to chilling at −10 °C for 5 min, from 18% to 65% (McLellan *et al.*, 1984). Many microbial cells are able to maintain their membrane fluidity at low temperatures by changes in the fatty acid composition of the lipid bilayer. The most common response is an increase in desaturase activity which decreases the proportion of saturated versus unsaturated lipid acyl chains in the bilayer. Other responses such as a shortening of acyl chain length, an increase in the extent of branching, or a decrease in the proportion of cyclic fatty acids can also occur and all have a similar effect in promoting membrane fluidity (Russell, 1984).

Certain microalgae in Antarctica have an unusually high microtubule stability even at sub-zero temperatures. These proteins are the structural elements of flagella, and one test of their stability is whether the cells are able to swim. *Dunaliella* sp. and *Chlamydomonas* sp. isolated from hypersaline lakes in the Vestfold Hills remained motile at temperatures as low as −14 °C. The swimming speed of *Pyramimonas gelidicola* from the same habitat dropped more or less exponentially over the temperature range 3−−9 °C. This alga ceased to swim at −10 °C, but even at −14 °C its flagella continued to beat. Microtubules also maintain the characteristic lobed, pyramidal shape of *Pyramimonas*, and the persistence of this morphology at these low temperatures provided further evidence of microtubule stability (Burch & Marchant, 1983). In contrast, the microtubule protein tubulin in some cold-adapted algal species may be unstable at moderately warm temperatures. Certain snow algae, for example, lose their flagella at temperatures above 4 °C (Hoham, 1975).

Microbial species inhabiting low temperature polar environments may have cold-adapted enzymes manifested as low apparent activation energies. Smith & Platt (1985) tested this hypothesis for RuBPCase in Arctic phytoplankton and found that contrary to expectation the apparent $E_a$ was higher than for the same enzyme in tropical populations. Their findings imply that factors other than temperature control the physiological differences between these two populations.

### 10.2.5  *Cellular freezing*

During the freeze–thaw cycles which characterise many antarctic environments microbial cells experience the following sequence of physical and chemical events (Mazur, 1969, 1970):

1. The cells and the external environment initially cool below their theoretical freezing points, a process known as *supercooling*. Ice soon begins to form externally but the cell wall and plasma membrane prevent these ice crystals from seeding the cell interior which remains supercooled and ice-free. Many plant, animal and microbial cells can be supercooled to at least −10 °C without freezing, perhaps because any aqueous channels through the membrane will be extremely small and will restrict the ice crystals growing into the cell to sizes that have very low melting points. Ice crystals 300 nm in radius, for example, will melt at −10 °C.

2. With a further decrease in temperature more of the extracellular solution will be converted to ice and the external concentration of solutes will increase. This will cause an osmotic pressure gradient across the plasma membrane, and to maintain equilibrium water will flow out of the

cell and freeze externally. The rate of dehydration by this process will be:

$$-dV/dt = LpA (\pi_o - \pi_i)$$

where 
$V$ = volume of cell water
$t$ = time
$Lp$ = permeability of the cell to water
$A$ = area of the cell surface
$\pi_o$ = osmotic pressure outside the cell, and
$\pi_i$ = osmotic pressure inside the cell.

Providing the cooling process is slow enough and the membrane is sufficiently permeable to water the cell will gradually desiccate and ice-crystal formation will be avoided. However, if $Lp$ is low or the cell is cooled too rapidly equilibrium will be established by internal freezing.

Both ice formation and dehydration can result in considerable, and potentially lethal, cellular damage. The loss of turgor pressure associated with the water efflux will cause a reduction in cell volume which may lead to additional injury. For some cells this freeze-induced contraction alters the resilience of the plasma membrane, and the cells burst during rewarming (Steponkus, 1981).

3. As ice formation continues the concentration of solutes both inside and outside the cell will increase, and certain compounds will eventually exceed their solubility and precipitate out. This process may also be accompanied by large changes in pH. The cells thus experience an unusual intracellular and extracellular environment. This solute concentration is considered another major factor contributing towards freezing injury (Steponkus, 1981). The release of Coupling Factor I (one of the proteins involved in generating ATP during photosynthesis) from thylakoid membranes during freezing may in part be due to these salt effects.

4. Eventually all free water in the system will be converted to ice, and all solutes precipitated. The temperature at which this first occurs is called the *eutectic* point or zone.

5. The ice may initially form as small, non-spherical crystals but these have a relatively high surface free-energy by comparison with large spherical crystals. The ice will therefore be gradually converted to the larger more stable crystalline forms. This process is known as *recrystallisation* (or grain growth) and can occur over a wide range of temperatures, even below the eutectic point. These larger ice crystals are likely to have more disruptive effects on cellular structure than smaller grains. The formation of intracellular ice appears to be lethal to all but a handful of microbial species (Smith *et al.*, 1986).

6. The reverse sequence will be followed during warming. The external medium will begin to melt once temperatures rise above the eutectic point. The external vapour pressure will then rise above that of the cells, and water will re-enter the cells at rates dependent upon the rate of warming and the permeability of the cellular membrane. Slow warming favours recrystallisation, and the resultant large ice crystals may damage the cell.

The ability of any microbial population to withstand freezing will therefore depend upon the rate of cooling, the minimum temperature reached, and the subsequent rate of warming. Slow cooling minimises intracellular ice formation, while fast warming minimises recrystallisation, and so this combination generally permits the highest survival rates (Fig. 10.4). A sudden drop in temperature – for example, due to an abrupt increase in wind velocity – may have the most deleterious effects on antarctic micro-organisms by encouraging cellular freezing. Very slow cooling rates, however, prolong the exposure of cells to unusual pH and ionic compositions that may further reduce survival rates. Microbial populations may therefore show an optimum cooling velocity for survival; for example the yeast *Saccharomyces cerevisiae* (Fig. 10.4).

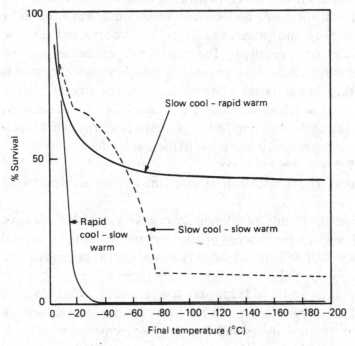

Fig. 10.4. The influence of cooling and rewarming rates on freezing survival by the yeast *Saccharomyces cerevisiae*. Redrawn from Mazur (1969).

### 10.2.6 *Adaptation to freezing*

Micro-organisms show large differences in their ability to withstand freezing, and a number of adaptive strategies may minimise or eliminate ice damage during the freezing process:

(i) Production of extracellular substances that reduce or prevent ice nucleation in the immediate vicinity of the cell. The mucilages which coat antarctic stream, soil and lake biofilms of cyanobacteria and certain diatom assemblages (e.g. in the sea ice) perhaps confer this protection.

(ii) Increased solute content that depresses the freezing point of the intracellular fluid. This physiological attribute also reduces dehydration by reducing the amount of water that needs to be removed from the cell to achieve osmotic equilibrium; a doubling of internal solute concentration, for example, will halve the extent of water loss at a given temperature during extracellular freezing. High solute levels will also allow non-specific colligative dilution of any toxic compounds which could potentially accumulate during the dehydration process. Sugars and sugar alcohols (polyols) appear to be especially effective protecting agents and accumulate to high concentration in certain organisms such as antarctic mosses and lichens (see Chapter 8).

(iii) Any process which slows freezing may enhance survival. The high heat capacities of water, and to a lesser extent rock and soil (Appendix 2) will dampen any rapid changes of temperature in these environments.

(iv) Increase in membrane water permeability ($Lp$ above) which will allow the cell to rapidly lose water to the sites of extracellular nucleation.

(v) Changes in membrane elasticity making the cell more resistant to the contraction and expansion stresses induced by water loss and rehydration. Plant protoplasts acclimated to freezing have a three-fold greater 'tolerable surface area increment' than non-acclimated protoplasts, and there are functional and structural differences between the two types of membrane. A detailed examination of algal cell membranes by electron microscopy has revealed an intriguing correlation between freezing tolerance and ultrastructure of the plasmalemma (Clarke & Leeson, 1985). Three species of snow algae, including an isolate of *Chloromonas palmelloides* from Signy Island, showed a tolerance to freezing and thawing down to $-20\,°C$; in all three isolates there were widespread invagi-

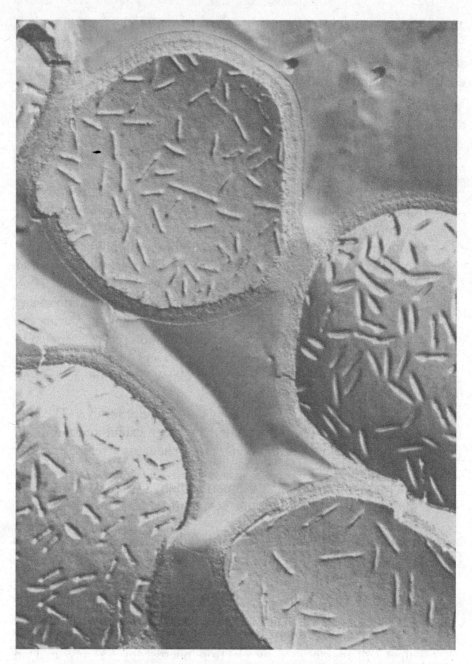

Fig. 10.5. Invaginations of the plasmalemma of a freeze-tolerant yeast, *Saccharomyces cerevisiae*. The invaginations are 0.1–0.8 μm long (typically 0.4 μm), and about 0.05 μm wide and deep. Similar ultrastructural features have been observed on the plasmalemma of freeze-tolerant snow algae from Signy Island. Freeze-fracture electron micrograph from Clarke & Leeson (1985), reproduced by permission of the authors and *Protoplasma*.

nations over the outer membrane. By contrast, two isolates of *Chlamydomonas* from much warmer habitats had no invaginations and a much poorer survival rate after freezing. Similar invaginations have also been observed in freeze-tolerant yeast cells (Fig. 10.5). The functional significance of these features has not yet been established, but it is possible that they might allow the cells to more effectively cope with the contraction and re-expansion stresses that accompany freezing.

(vi) Certain microbial species survive the extreme winter temperatures of Antarctica as highly resistant spores. The brightly-coloured zygotes of the snow algae, for example, probably have a much greater freezing tolerance than the motile vegetative cells. For some bacterial species the process of endospore formation partially dehydrates the cell which renders it unfreezable. During spore development water molecules move from the main body of the cell where freezing would be highly deleterious to outer peptidoglycan polymers where the water is held by $COO^-$ groups. The cells at this time may also produce trehalose, a disaccharide that is known to protect proteins against denaturation and phospholipid membranes against phase transition (Franks, 1985).

## 10.3    High salinity

In many of the microbial habitats of Antarctica dissolved ions and other solutes have been greatly concentrated by freezing and evaporation. Hypersaline lakes and soils occur throughout the region. The sea-ice biota experience wide fluctuations in salinity of the water draining through brine channels in the ice. Even ice-shelf environments experience saline conditions where seawater and other brines have been freeze-concentrated. Mirabilite deposits have been recorded on the McMurdo Ice Shelf, for example, where some of the partially unfrozen pools attain salinities up to 100‰. In certain antarctic habitats, such as coastal ponds, the microbial flora must sustain not only the extremes of salinity, but also large variations in salt content over relatively short periods of time.

The primary influence of a high solute content in the aqueous environment is to lower the freezing point of water (Section 7.2.2). The effect has far-reaching implications for the survival and activity of antarctic microorganisms. Hypersaline lakes and pools remain ice-free and mixing during winter at several locations on the continent, and provide a liquid but extremely cold environment for microbial growth. The micro-organisms in Deep Lake, Vestfold Hills experience winter temperatures below −30 °C.

A less extreme example is found in the southern parts of the Southern Ocean where water temperatures just below 0 °C persist throughout the year.

Hypersaline soil conditions extend the period of thaw by bringing forward the first date of melt into cooler conditions, and prolonging the late summer growth period before refreezing sets in. Hypersaline environments will be less susceptible to diel freezing and the associated damaging effects of rapid dehydration and cellular ice crystal formation (see above). For those organisms which can withstand the adverse effects of high salt concentrations, hypersaline lakes and soils may offer a more persistent aqueous environment.

Hypersalinity has a wide range of other less obvious effects. Hypersaline groundwaters continue to flow during the antarctic winter, and saline streamwaters may have an unusually prolonged season of discharge. Salinity gradients in lakes inhibit wind-induced mixing, although thermo-haline convection cells may be established at moderate salinities. The stable gradients allow a slow accumulation of heat, and several antarctic lakes demonstrate an inverse temperature structure, with deep strata that remain at temperatures well above freezing throughout the year. Highly saline brines may also form within sea ice. This dense, cold liquid can then drain down through the ice and may have a toxic effect on the microbial communities that it comes in contact with. Palmisano, Soo Hoo & Sullivan (1987b) found that a 60‰ solution of brine collected from the upper layers of congelation ice in McMurdo Sound completely inhibited photosynthesis by the sea-ice microalgae. In this study higher photosynthetic rates were recorded at 20‰ than at 33‰, and in earlier work by Bunt (1964) optima for the platelet community were recorded at 7.5–10‰. Palmisano *et al.* (1987b) suggest that microalgal growth may be confined to the lowermost 5–20 cm of the ice profile because this region is more frequently flushed and diluted with the underlying seawater. Higher in the profile salinities may reach more than 100‰, which would completely suppress algal metabolism and growth.

At a cellular level high salinities impose an effect through at least three general mechanisms:

   (i) Dehydration because of the high osmolarity of the medium.

   (ii) Changes in the water activity of both the medium and the cell.

   (iii) Direct influences of the concentrated solutes themselves.

The primary adaptation of microbial cells to minimise these effects is the production of compatible solutes which maintain a high internal osmolarity without impairing enzyme activity and other cellular functions. *Dunaliella*, a phytoflagellate genus that is represented in several hypersaline lakes in Antarctica, produces glycerol as its primary solute (e.g. Ben-Amotz, 1975). Many halophilic bacteria, on the other hand, accumulate potassium chloride which is less inhibitory than sodium chloride (Kushner, 1978). The enzymes from some of these bacteria although salt-tolerant seem to be sensitive to low temperatures. However, several halotolerant bacteria have been isolated from Antarctica that can grow under the double extreme of cold, highly saline conditions.

*Planococcus* isolated from soils in the Dry Valleys was capable of growth at 2.0 M NaCl even at a temperature of 0 °C (Miller, *et al.*, 1983). This isolate had no specific NaCl requirement but rather growth rates declined with increasing salt concentration. A variety of other cocci as well as motile

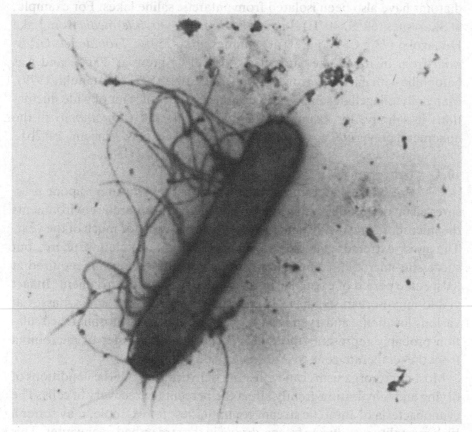

Fig. 10.6. *Halomonas subglaciescola*, a halotolerant bacterium from a Vestfold Hill lake. The cell is about 1 μm wide and 4 μm long, with flagella and thick appendages. Shadow electron micrograph from Franzmann *et al.* (1987b), reproduced by permission of the authors and the *International Journal of Systematic Bacteriology*.

and non-motile rods were also isolated from the soil during this study using saline (0.85 M NaCl, 0.2 M MgSO$_4$) enrichment media. All of these organisms tolerated up to 2 M NaCl and had temperature optima in the range 15–25 °C. The halotolerant *Planococcus* strain A4a showed a marked increase in total intracellular amino acid concentration with increasing salinity, but no response in cellular potassium levels. In this species free amino acids, not potassium, may act as compatible solutes (Miller & Leschine, 1984).

Another bacterial species with a high level of salt tolerance has been isolated from a lake in the Vestfold Hills. The isolate, named *Halomonas subglaciescola* (Fig. 10.6) grew over the temperature range −5–+25 °C at salinities up to 200‰ (Franzmann *et al.*, 1987b). Various salt-tolerant diatoms have also been isolated from antarctic saline lakes. For example, at Skarvsnes (69 °S, 40 °E) *Achnanthes brevipes* var. *intermedia* from Lake Hunazoko (73 g Cl l$^{-1}$) grew at salinities up to 250‰. *Tropidoneis laevissima* from nearby Lake Suribati (25 g Cl l$^{-1}$) grew at 200‰, and was halophilic with maximal growth in the range 100–150‰ (Watanuki, 1979). Many phytoflagellates in Antarctica appear to be tolerant of wide fluctuations in salinity, for example *Chlamydomonas* and *Chroomonas* in the ephemeral ponds at Cape Bird, Ross Island (Vincent & Vincent, 1982b).

## 10.4    Dehydration

Freezing temperatures, high salinity and low water vapour pressures all favour cellular dehydration, and in several antarctic environments the microflora must withstand prolonged desiccation for much of the year. The most extreme example is probably the antarctic atmosphere, but successful microbial colonisation throughout the region has required at least some period of existence in this cold, dry aerial environment. In fact viable micro-organisms have been isolated from antarctic air masses at various locations, and even with the advent of man, atmospheric circulation probably represents the main mechanism of transfer of microbiota from the temperate zone to Antarctica (see Chapter 11).

Most micro-organisms can withstand dehydration, but the conditions of drying and rehydration greatly affect the percentage recovery of cells. The cyanobacteria of antarctic stream communities, for example, may suffer a high mortality rate during freeze-drying in the streambed over winter. This may result in high standing stocks of non-living carbon and chlorophyll *a* associated with damaged cells that are no longer viable. The photosynthetic rates that have been measured in these communities when the streams are flowing are very low when expressed per unit carbon or per unit

a

b

Fig. 10.7. Cyanobacteria can remain dry, frozen and viable throughout the antarctic winter, and then resume photosynthesis almost immediately upon rehydration. (*a*) The investigator is holding a piece of freeze-dried mat from the extensive sheets of *Nostoc commune* that overwinter in Fryxell Stream, Dry Valleys. (*b*) Extensive black crusts of *Calothrix* and *Gloeocapsa* coating the dry bed of a stream in the southern Dry Valleys (Upper Trough Stream) about three weeks prior to streamflow.

chlorophyll *a* (see section 7.4.1), probably reflecting this large component of killed or freeze-damaged cells (Vincent & Howard-Williams, 1986, 1988). *Nostoc commune* is a common dominant in stream and moist soil communities in Antarctica (Fig. 10.7) and temperate latitude isolates of this species are known to metabolically respond to rewetting very rapidly even after prolonged desiccation (e.g. Potts & Morrison, 1986).

In general cyanobacteria and chlorococcalean green algae seem to be especially tolerant of desiccation (Fogg, 1969). Both groups are the phototrophic dominants in hot, arid deserts, and similarly predominate in the cold, arid environment of Antarctica. Antarctic cyanobacteria and green algae may have a greater tolerance for freeze-drying than their temperate latitude counterparts. For example, terrestrial and freshwater isolates from both groups in the McMurdo Sound region showed a much greater resilience to freeze–thaw cycles than a range of isolates from Wisconsin, USA (Holm-Hansen, 1963).

Mucilage production seems to be often correlated with drought resistance, possibly because this material slows the rate of water loss from the cells during desiccation (Shephard, 1987) and prevents ice formation in the immediate vicinity of the cells. The success of mucilaginous cyanobacteria throughout Antarctica may rest upon their ability to survive the harsh combination of low temperatures, freezing, desiccation and periodic exposure to elevated solute concentrations during ice formation. Relative to eukaryotic algae this microbial group seems to have an 'outstanding capacity' to endure such a combination of adverse extremes (Fogg, 1969).

## 10.5    Low light

Persistent ice cover is a prominent feature of many inland water habitats and marine environments throughout the Antarctic. Periodic or prolonged snow cover further coats the ice and also the terrestrial catchments through at least part of the year. Soil- and rock-dwelling microbes often exist well below the surface to escape the effects of dehydration and freeze–thaw cycles. The phototrophs in all of these ecosystems are shaded from direct sunlight and experience a low and often limiting availability of radiation for photosynthesis.

Shade-adapted populations are well known from many of these energy-limited ecosystems. The mechanisms of adaptation include:

   (i) *Saturation of photosynthesis at low intensities of photosynthetically available radiation (PAR).* The sponge spicule diatom communities in McMurdo Sound saturate at <20 $\mu$E m$^{-2}$ s$^{-1}$ and low saturation intensities similarly characterise ice-covered lakes, the

Table 10.2. *Photosynthesis–light responses by phototrophic communities from various antarctic environments*

$P_{max}$ is the photosynthetic rate at light saturation (mg C (mg chlorophyll $a$)$^{-1}$ h$^{-1}$), $\alpha$ is the slope of the light-limited portion of the curve (mg C (mg chlorophyll $a$)$^{-1}$ h$^{-1}$ ($\mu$E m$^{-2}$ s$^{-1}$)$^{-1}$, $I_k$ is the irradiance at the intersection between $\alpha$ and $P_{max}$ ($\mu$E m$^{-2}$ s$^{-1}$), and photoinhibition refers to the maximum irradiance ($I_m$).

| Community/site | $P_{max}$ | $\alpha$ | $I_k$ | $I_m$ | Photoinhibition | Reference |
|---|---|---|---|---|---|---|
| *Sea ice* | | | | | | |
| Bottom congelation ice (−2°C) | 0.105 | 0.014 | 7.4 | 506 | strong (c. 70%) | Palmisano *et al.* (1987b) |
| Platelet ice (−2°C) | 0.320 | 0.027 | 11.7 | 506 | no | Palmisano *et al.* (1987b) |
| Surface pool (tide crack) | 3.94 | — | — | 1600 | | Palmisano & Sullivan (1985b) |
| *Marginal ice zone* | | | | | | |
| 9–10-tenths ice cover | 1.3 | 0.012 | 150 | 1000 | no | Marra & Boardman (1984) |
| 6-tenths ice cover | 3.8 | 0.027 | 150 | 1000 | yes (c. 20%) | Marra & Boardman (1984) |
| *Open ocean* | | | | | | |
| Phytoplankton | 0.4–1.2 | — | 22–175 | 1500 | variable | Tilzer *et al.* (1985) |
| Phytoplankton | 0.7–4.4 | 0.01–0.05 | 100–200 | 1700 | yes (up to 50%) | Sakshaug & Holm-Hansen (1986) |
| *Marine benthos* | | | | | | |
| *Trachyneis aspera* | 0.2 | 0.013 | 20 | 300 | no | Palmisano *et al.* (1985c) |
| *Lakes* | | | | | | |
| Lake Fryxell | 0.1 | — | 2 | 10 | — | Vincent (1981) |
| Heywood Lake (Aug) | 0.4 | — | 11 | 120 | no | Hawes (1985) |
| *Cryptoendoliths* | | | | | | |
| Lipid biosynthesis | — | — | 200 | 1000 | no | Vestal (1985) |

bottom of the sea ice and deeply mixed marine algal communities (Table 10.2).

(ii)  *High cellular pigment concentration.* The light-capturing ability of antarctic shade organisms is manifested as a sharp gradient over the light-limited portion of the *P* vs *I* curve, but low rates of photosynthesis per unit pigment at $P_{max}$ (Table 10.2). $P_{max}$ in several communities shows a marked increase with improved irradiance conditions – for example in the marginal ice zone with increasing water column stability and decreasing ice cover, and in lakes with the breakup of their ice cover (Table 10.2).

(iii) *Small cells.* Package size has an important influence on the light-harvesting ability of pigments, and small cells more efficiently capture PAR than large cells. Small-celled algae, so-called picophytes, are found in various low-light environments in Antarctica, including ice-covered lakes and endolithic habitats. In these environments, however, a small cell-size may be selected for by other constraints – the lack of turbulence in ice-covered lakes, and the small volumes available for microbial colonisation and growth between rock crystals. Large diatoms are found at the bottom of the sea ice, in the deeply mixed open waters of the Southern Ocean and in the benthos (e.g. the long (up to 400 $\mu$m) pennate species *Trachyneis aspera* in McMurdo Sound, Table 10.2), and small cell-size is clearly not a prerequisite for growth under shade conditions.

## 10.6    High radiation

In summer certain antarctic microbial communities inhabit environments which receive continuously high intensities of solar radiation. Surface snow and ice environments with their high albedo provide extreme examples, but surface rock, soil, lakewater and marine communities may all experience bright light conditions at certain times of the year. High intensity PAR has an inhibitory effect on the photosynthetic apparatus, and at long time-scales may cause irreversible damage. Strong photo-inhibitory responses have been measured in two small ponds on Ross Island (Goldman, Mason & Wood, 1963). In these waters phytoplankton photosynthesis was completely out of phase with the 24 h radiation cycle with peak carbon fixation rates occurring near midnight when light availability was at a minimum.

The photochemical effects of continuous daylight over 24 h in summer have yet to be explored although again cyanobacteria seem to be especially resilient to this environmental extreme. Cells of *Anacystis nidulans*, for example, can remain intact and viable after extensive and prolonged photo-oxidation (Schmetterer & Pescheck, 1981) and in Antarctica many cyanobacterial species produce high concentrations of carotenoids that probably confer protection against bright light (e.g. Howard-Williams & Vincent, 1988). Ultraviolet radiation causes various cellular lesions but its most damaging effects are on nucleic acids. Recent evidence of a spring 'ozone hole' over parts of Antarctica (Farman, Gardiner & Shanklin, 1985) suggest that the ultraviolet dosage may be unusually high for some microbial populations, but the extent of any radiation effects is at present unknown.

## 10.7   Winter darkness

At latitudes higher than the Antarctic Circle, phototrophs must experience another adverse extreme, the continuous polar night for several weeks of the year (Appendix 2). Within the terrestrial ecosystems the cyanobacteria and other microbiota survive this period in a freeze-dried and metabolically inactive condition. In the lake and sea environments the continuous lack of an energy supply presumably causes large mortalities in the phototrophic populations, but various survival mechanisms may contribute towards the persistence of some cells that can then form the inoculum for the next season of growth. There is evidence that the uptake and utilisation of dissolved organic compounds may contribute toward the winter survival of sea-ice diatoms (Palmisano & Sullivan, 1983a). Many of the lake communities are dominated by phyto-flagellates, and representatives of this group are known to be capable of supplementing their carbon nutrition by ingesting small particles such as bacteria. It has not yet been demonstrated, however, that facultative phagotrophy in these ecosystems operates either as a winter survival mechanism or as a nutritional mode for active growth.

In both the marine and freshwater environments low temperature is probably a major factor contributing towards the overwinter survival of phototrophs. Respiration rates of phytoplankton in the Southern Ocean, for example, have measured $Q_{10}$ values ranging from 2.3 to 11.9 (Table 10.1), and therefore the cellular maintenance costs are greatly reduced at the low ambient seawater temperature (Tilzer & Dubinsky, 1987). Slow catabolic losses in this and other antarctic environments will contribute towards the long-term viability of cells experiencing short daylengths or prolonged darkness during periods outside the polar summer.

# 11

# Microbes and humans in Antarctica

## 11.1 Introduction

Human activities have become an important new factor in the antarctic environment (Fig. 11.1). Ice-free sites total a very small area and are major centres for biological activity, but they are preferred localities for research stations and scientific expeditions, and may be especially vulnerable to human disturbance. With the proliferation of large national bases in the second half of this century substantial quantities of materials not previously encountered in Antarctica have been imported into the region. Some of these, such as petroleum fuels and drilling fluids, are likely to cause a radical shift in the local distribution and abundance of native microbiota. Similarly, compounds such as the inorganic and organic nutrients contained within human wastes are likely to stimulate certain elements of the microflora, but at the expense of the original community structure and properties of the environment. Micro-organisms under water stress such as in arid soils, saline lakes or areas exposed to continuous freezing and thawing may be particularly susceptible to these increasing quantities of pollutants now entering the Antarctic.

Microbes from the temperate zone have undoubtedly accompanied the human settlement of Antarctica, but it is not known whether these introduced species have the capacity to displace or modify the existing assemblages. The introduced species are likely to be most successful (and also introduced in large numbers) in environments polluted with substrates from the temperate zone. However, a large and persistent inoculum of alien species might also influence the community structure of the native microbiota in otherwise unperturbed habitats. Genetic engineers are increasingly searching the microbial flora of extreme environments as a source of new enzymes and other biochemicals that can be used for industrial or pharmaceutical purposes (e.g. Daniel, Morgan & Hudson,

Fig. 11.1. Establishment of the Greenpeace antarctic base at Cape Evans, Ross Island. In the background to the right is the hut which was the base for Scott's final expedition (1910–13). Photography by A. Loor, Greenpeace, 1987, reproduced by permission.

1987). An actinomycete isolate from Mt Erebus, Ross Island with promising antibiotic properties (Greenfield, 1983) suggests a biochemical potential of direct utility to man. Antarctica supports a wide range of microorganisms that seem highly resilient to the major chemical changes associated with freezing. Such resilience is especially attractive for industrial applications. These observations imply that there are strong commercial as well as environmental grounds for protecting key antarctic habitats from future shifts in microbial diversity.

The antarctic environment offers other microbiological opportunities of interest to man. Most of the research settlements in Antarctica are isolated from each other and the rest of the world for several months of the year. These semi-closed, well-defined human communities present ideal experimental systems in which to investigate the transmission and spread of pathogens. Medical researchers have already used this opportunity to gain new insights into the epidemiology and control of the common cold.

## 11.2    Medical microbiology

Research stations offer a peculiar antarctic habitat for microbial growth. They are continuously warm and illuminated, but partially retain the dry air characteristics of the external atmosphere (Table 11.1). These island-like habitats are isolated from other settlements for long periods of the year and present well-defined human communities in which to study the dissemination of certain microbes of medical concern. To date most of this type of research has focussed on the viruses responsible for respiratory tract illnesses. These studies have provided new information about the aetiology of the common cold, as well as an opportunity to test a new antiviral product.

The records of isolated field parties and bases in Antarctica have supported the belief that the incidence of respiratory illness (common colds) becomes greatly reduced within a few weeks of isolation. The overwintering personnel at McMurdo Station and adjacent Scott Base, for example, experience a low incidence of colds, but the infection rate abruptly rises at the winter-fly-in period (WINFLY) in late August when a substantial number of personnel are flown in from New Zealand. The community then remains isolated for five or more weeks until flights resume at the start of the summer field season. This closed community of 150 or more people in which most of the viruses present have been brought in on the WINFLY flights have offered an ideal experimental opportunity to examine the transmission of respiratory viruses.

Two incoming Scott Base personnel in WINFLY 1977 had moderate colds and yielded two different rhinoviruses. Serum antibody tests showed

Table 11.1. *Environmental conditions at South Pole Station*

| Location | Temperature (°C) | Relative humidity (%) | Water vapour density (g m$^{-3}$) |
|---|---|---|---|
| *Biomedical building* | 21.4 | 8 | 1.5 |
| *Science building* | 19.8 | 33 | 5.9 |
| *Outside the buildings*[a] | | | |
| January | −26.9[b] | 100 | 0.46 |
| July | −61.0 | 100 | 0.01 |

[a] but beneath a protective dome that covers the station buildings
[b] monthly mean temperatures
From Parkinson *et al.*, 1983.

that 13 of the 22-person base staff were susceptible to these viruses, yet despite many hours of close contact the two infected individuals did not transmit their viruses to anyone. In the same period there was an outbreak of an adenovirus disease at McMurdo Station. Of the 200 station personnel, 88% seemed to be free of the serum neutralising antibody, and thus susceptible to the virus. The attack rate in susceptibles was only 14.5% and the disease spread slowly with three to five new cases per week. No one at Scott Base was infected, and viral transmission seemed to require prolonged close contact. Over the WINFLY periods of three separate years the dissemination of cold virus was slow and the epidemic curves were relatively flat. These observations were consistent with volunteer experiments in North America, and show that viral infections of the respiratory tract are surprisingly difficult to transmit (Dick *et al.*, 1978).

The WINFLY community was subsequently used to test a virucidal paper handkerchief (VPH) that was impregnated with iodine. In laboratory trials small pieces of VPH were able to inactivate strong titres of rhinoviruses within a few minutes, and the infectivity of viral suspensions placed on the skin could be reduced by a factor of 100 000 over a similar period of time (Dick *et al.*, 1979). Field applications both during WINFLY and subsequently during the summer period at Scott Base indicated that VPH usage could reduce the incidence of colds by more than 50% (Dick *et al.*, 1980; Meschievitz *et al.*, 1982).

Although respiratory disease becomes very rare beyond the first six weeks of isolation, periodic infections may still occur. An outbreak of common colds has been documented at the British antarctic base on Adelaide Island some 17 weeks after complete isolation (Allen 1973). Episodes of respiratory infections have also been regularly observed during the 8.5 month isolation period at South Pole Station (Parkinson *et al.*, 1983).

A possible explanation for these outbreaks of infection is that the cold, dry antarctic environment (Table 11.1; see also Appendix 2) prolongs the survival of the causative virus; human parainfluenza viruses are known to remain infective for longer at low relative humidities, and they may be stored below $-60\,°C$ for several years without loss of infectivity. This hypothesis was tested by exposing viral samples sprayed onto petri plates to indoor ($21\,°C$, water vapour density of $1.5$ g m$^{-3}$) and outdoor ($-22$ to $-33\,°C$, $0.71$–$0.25$ g m$^{-3}$) environments at the South Pole. Parainfluenza virus type 1 was inactivated after four days at room temperature, and after seven days outside. Parainfluenza virus types 2 and 3 were inactivated after seven and 12 days, respectively, at room temperature, and after 17 days of storage outside. Thus the long-term survival of parainfluenza virus in either environment is highly unlikely. It is more probable that the viruses responsible for colds during prolonged isolation are shed from one or more humans with a persistent viral infection in their upper respiratory tract. The cold, dry air has been shown to impair nasal ciliary activity and thereby reduce or eliminate the distribution of antiviral substances. This effect would enhance the production of virus in persistently infected individuals, as well as increase the susceptibility of others to viral infection (Parkinson *et al.*, 1983).

Other micro-organisms of medical interest have received much less attention in Antarctica. Human and animal wastes can persist undegraded in terrestrial parts of Antarctica, and legacies from earlier field camps have the potential to contaminate the local environment with pathogenic micro-organisms for prolonged periods of time. Viable *Escherichia coli* cells have been isolated, for example, from 1907 expedition pony dung at Cape Royds more than 50 years later (Boyd & Boyd, 1963). These isolates showed a greater survival rate in field tests than recently introduced strains of *E. coli*, suggesting a gradual selection by the environment for hardier genotypes. Surface contamination of snow and soils with *E. coli* has also been demonstrated near research stations (e.g. Cameron, 1972b; Zunino, Castrelos & Margni, 1985). The coastal stations in Antarctica typically discharge their sewage into the sea, but although the ocean has a vast buffering capacity for such contaminants the local influence of human pathogens on the marine birdlife and mammals is unknown.

## 11.3    Pollutants and microbial responses

As scientists and explorers we have brought with us an enormous array of new substances into the antarctic environment. Many of these introduced materials have the potential to radically alter the physical and chemical properties of microbial ecosystems. To date, most of the pollutants have entered the environment during scientific research programmes. For example diesel fuel Arctic-grade (DFA) and calcium chloride solutions have been commonly used to facilitate investigative geological drilling through permafrost. Spills and leaks of these drilling fluids have been inevitable and frequent, although generally over very small areas.

DFA-saturated soil in the Dry Valleys appears to persist for many years and forces a shift away from the native assemblage to unusual species, either native or introduced, that are capable of tolerating or degrading hydrocarbons. Soils at New Harbor contaminated with DFA have shown 100-fold increases in the number of colony-forming organisms. Soils near

Lake Fryxell contaminated with calcium chloride drilling muds showed more than a factor of ten decrease in number of colony-forming microbes, and a drastic reduction in diversity to a single bacterial species (Parker & Howard, 1977). Experimental application of crude oil to a small area of soil at Cape Bird, Ross Island, similarly resulted in a marked drop in microbial species diversity (Konlechner, 1985). The assemblage shifted towards hydrocarbon-degrading bacteria, with an increase in total plate counts. Mineralisation of the organic nitrogen fraction in the ornithogenic soils was inhibited by the crude oil for up to two years following the experimental spillage.

An 'environmental impact matrix' has been devised as a way of expressing the impacts of drilling in Antarctica on native soil, water and air microbiota (Table 11.2). This approach incorporates an assessment of the relative magnitude of the environmental impact (0 = no impact; 4 = complete destruction of habitat) at time-scales from diurnal periods to greater than 15 years (Parker & Howard, 1977). Subjective decisions are

Table 11.2. *An environmental impact matrix for drilling on Ross Island*

Each effect is rated on a scale from 0 (no impact) to 4 (complete destruction).
In each group of five digits the first number refers to the impact over the first day,
the second to impacts at the seasonal level, the third to the 1–5 year impact,
the fourth to the 5–15 year impact, and the fifth to impacts at time scales longer
than 15 years.

| Drill site activity | Type of impact | | |
| --- | --- | --- | --- |
| | Changes in soil properties (subsurface) | Changes in the soil surface environment | Changes in the indigenous microbiota |
| Well drilling | 11000 | 33211 | 11111 |
| Drainage modification | 12344 | 11111 | 33211 |
| Drilling fluid spills | 43321 | 33333 | 43211 |
| Disposal of sewage and other wastes | 44444 | 33333 | 44444 |
| Introduction of alien micro-organisms | 00000 | 10000 | 44433 |
| Disposal of tailings | 33333 | 11110 | 33211 |
| Vehicle movements | 11111 | 32211 | 42111 |

Modified from Parker & Howard, 1977.

still required to code a particular impact, but this analytical technique offers a first step towards quantifying the overall influence of human activities on antarctic ecosystems. Further criteria for environmental impact analysis in Antarctica have been compiled by the Scientific Committee on Antarctic Research (Benninghoff & Bonner, 1984).

The impact of research activities in Antarctica is still small and localised (but significant, see Cameron, 1972b; Fifield, 1985) relative to the potential effects of future tourism and resource exploitation through the region. A major oil spill during drilling operations in the Ross Sea, for example, would have devastating effects on marine microbes as well as organisms at other trophic levels (see Keys, 1984). Hydrocarbon-decomposing micro-organisms seem to be present in Antarctica, but as in the Arctic (Atlas, 1986) their rates of activity are likely to be extremely slow, and any oil spills would persist for long periods of time.

The solid phase of certain antarctic environments has already been modified by man but the microbial response to these changes has been little explored. The surface structure of arid soils such as in the Dry Valleys is easily disrupted by vehicles and footsteps, and these disturbances may remain visible for more than a decade (Cameron, 1972b). Large quantities of particulate material are released from antarctic bases and dispersed by wind, both during and after construction. The glass fibres from insulation material may be found in nearby soils to a depth of several centimetres. Snow and dust traps installed near Scott Base and McMurdo Station on Ross Island collected a range of particulates, but the most common organic materials were single cells and tissue fragments of conifer wood, probably from packing cases. The next most common particles were natural and synthetic particles from clothing (Benninghoff & Benninghoff, 1985). Pieces of wood, metal, plastic and other materials from abandoned campsites are commonly encountered in the Dry Valleys. Floating litter including polystyrene and other artificial materials is now distributed throughout the Southern Ocean. From their limited series of tows in one sector of the ocean Gregory, Kirk & Mabin (1984) estimate that about 250 million fragments of plastic and man-made items may be afloat in the sea between latitudes 45 and 65 °S. This concentration of material is negligible relative to the total area considered, however small plastic particles have been increasingly found in the digestive tract of antarctic seabirds (Benninghoff & Bonner, 1984).

## 11.4    Eutrophication in the Antarctic

Some antarctic environments contain sparse concentrations of nutrients and support a low biomass community of micro-organisms adapted to these oligotrophic conditions. The soils of the Dry Valleys, for example, failed to yield many microbial species until a low nutrient enrichment medium was developed (Vishniac, 1983). Various lakes and streams throughout the region receive low inputs of nitrogen or phosphorus and their phototrophs and associated communities are severely restricted in population size. Lake Vanda (Dry Valleys), contains $<0.1$ mg chl$a$ m$^{-3}$ at most depths throughout its euphotic zone and has been identified as one of the world's most oligotrophic natural waters (Goldman, Mason & Hobbie, 1967; Vincent & Vincent, 1982a).

All of these soil and water systems are highly responsive to any enhancement of nutrient supply. The surface waters of Lake Vanda, for example, contain $<1$ ppb (mg m$^{-3}$) of dissolved inorganic phosphorus, and a relatively low level of catchment development and input of human wastes to this system could dramatically increase algal growth. This would induce a deeper zone of anoxic bottom waters and would probably alter the species composition and diversity of planktonic microbes.

## 11.5   Microbial dispersal

With the advent of increasingly frequent aircraft and shipping movements between the temperate latitudes and many sites throughout Antarctica, mankind has become a potentially important agent of dispersal. Microbial propagules may now enter relatively benign environments such as permanent lakes and geothermally heated ground within hours to days of leaving the temperate zone. These micro-organisms carried on clothing or expedition cargo will not necessarily experience freezing and drying before they encounter favourable antarctic habitats for growth. This method of transfer thereby contrasts with the primary dispersal mechanisms which operate throughout the region, and may ultimately lead to a change in species composition or diversity in certain environments.

Most non-marine microbial species have probably entered Antarctica as dry, frozen propagules blown in by the wind. Many micro-organisms, even thermophilic species, retain their viability despite freezing and dehydration. Viable algae, particularly soil representatives, have been reported in the 'aerial plankton' of the temperate zone at altitudes up to 2000 m (Brown, Larsen & Bold, 1964), and at least at a local scale these air-borne cells may be very important for the inoculation of certain habitats. On Signy Island, for example, Broady (1979c) recovered up to several hundred algal propagules m$^{-2}$ h$^{-1}$ on nutrient agar plates exposed to the prevailing wind. Biogenic material in clouds over the Ross Ice Shelf included fibrous aggregations of algal cells, and infra-red absorption spectra of water samples from these clouds have consistently indicated the presence of proteinaceous material. These biogenic particles may initiate condensation and freezing and thereby help these clouds to precipitate (Saxena, 1982).

The origins of aerial microbiota in the Antarctic are not well defined, but dry algal mats may be an important source of propagules in some localities. Members of the Oscillatoriaceae, for example, survive repeated freeze-drying and rehydration and large areas of dry, brittle flakes and sheets of these cyanobacteria are encountered in parts of the McMurdo Ice Shelf. These must provide a large inoculum for new and existing lakes and ponds, and may overwhelm any chance introductions from the temperate zone. This 'ice shelf microflora' probably endured the periods of glacial advance in cryoconite habitats (see Section 2.3.2), to perhaps subsequently dominate the microbial inoculum in air masses which blow across the non-marine Antarctic today. Saxena's (1982) study identified the algae in Ross Ice Shelf clouds as *Planktonema lauterbornii*, a relatively rare freshwater species that has not been reported in the southern Victoria Land flora (Seaburg *et al.*, 1979). Scanning electron micrographs of these algae bear at least a superficial resemblance to oscillatoriacean flakes, but the original identification cannot be confirmed without light microscope observations.

Another mode of microbial transfer into Antarctica is on the outside of migrant birds. Specimens of south polar skua (*Catharacta maccormicki*), southern black-backed gull (*Larus dominicanus*), arctic tern (*Sterna paradisaea*), cape pigeon (*Daption capense*) and giant fulmar (*Macronectes giganteus*) were collected live at seven sites from 59 to 67 °S and 43 to 64 °W. They yielded a wide range of viable algae and protozoa. For example, cultures from sterile washings of the tail feathers from an arctic tern specimen yielded the cyanobacteria *Schizothrix calcicola* and *Nostoc commune*, the green alga *Chlorella vulgaris*, a *Bodo*-like protozoan, and two unidentified zooflagellates. Many of these bird species have very wide geographical ranges. South polar skuas banded at Palmer Station (65 °S), for example, have been recovered from Argentina, Mexico and Greenland, and both this species and the arctic tern may transport microorganisms between polar regions (Schlichting, Speziale & Zink, 1978).

The present magnitude and success of alien introductions into Antarctica is completely unknown, although the presence of exotic microbial species has been frequently recorded. Thermophilic bacteria and common fungi from the temperate zone have often been detected in microbiological surveys near research stations and field camps, and even highly experienced ecologists have noted how difficult it is to visit an environment without accidentally introducing alien microbiota:

At Mt. Howe (87°21'S, 149°18'W) during the 1970–71 field season, it was unfortunate that (an exotic) fungus was inadvertently carried to the farthest south exposed area of rock and soil. Upon opening a can of C-rations, it was found contaminated with a luxuriant growth of *Penicillium*. The wind blowing into the tent subsequently spread the fungus and contaminated the local area as shown in our next sample collection (Cameron, 1972b).

Although there have been frequent calls for the organised collection of baseline data for antarctic ecosystems (e.g. Paterson & Knox, 1972; Rudolf & Benninghoff, 1977) the assay technology to adequately monitor microbial communities has not existed. It is difficult to reliably collect and then interpret simple plate count data. Alien species are easily introduced with the investigation team and their equipment, and as illustrated by the Dry Valley soil yeasts, the native microbiota may have unusual growth requirements that are not met in a standard assay. Microbial lipid profiling now offers a powerful new approach towards this general monitoring problem, and has been applied with success to experimental investigations of the marine benthos in McMurdo Sound (Smith *et al.*, 1986). Unlike traditional plate counts and identifications, the assays based on these microbial lipid 'signatures' generate both biomass as well as diversity information without the need for artificial enrichment.

## 11.6    Endemism and displacement

There undoubtedly has always been some exchange of microbial propagules between Antarctica and temperate latitudes by migrant birds and atmospheric processes. Nonetheless, the reduced range of microbial species in some habitats seem more a function of the biogeographical isolation of Antarctica than the absence of suitable opportunities for growth. The reports of endemic yeasts in the soils of the Dry Valleys suggest an evolutionary divergence of at least certain antarctic species from the temperate zone genotypes. Similarly, newly described taxa such as *Hemichloris antarctica*, a cryptoendolithic alga from rocks in the Dry Valleys; *Notodendrodes antarctikos*, a foraminifer living on the surface of the Ross Sea sediments; and *Halomonas subglaciescola*, a bacterium from a hypersaline lake in the Vestfold Hills as well as a wide range of marine and freshwater microalgae support the belief that certain genetically distinct species may be unique to Antarctica. The recent development of ribosomal-RNA sequencing techniques (e.g. Pace, Olsen & Woese, 1986) now offers a convincing test of microbial endemism that would be ideally applied to antarctic organisms.

Many of the microbial life-forms found in Antarctica do not seem to be especially adapted to these environments, but rather are simply cold-tolerant or desiccation-tolerant species that are also found in the temperate zone. At the community level, however, the microbiota is often composed of an unusual assemblage that reflects the pecularities of the biological environment as much as the physical and chemical extremes. The muted grazing pressure in many of the inland aquatic environments, for example, has allowed microbial assemblages to flourish that in the presence of temperate latitude biota such as aquatic arthropods might be quite differently structured. It is possible, for example, that certain crustacean zooplankton species from Signy Island (e.g. *Branchinecta*) or the Vestfold Hills (e.g. *Daphniopsis*) could survive and grow within Dry Valley lakes. The introduction of this new trophic level would have an unpredictable but potentially large impact on the Dry Valley lake ecosystems. The microbial assemblages at present experience almost negligible grazing pressure, and major changes in species composition might be expected with the introduction of herbivores. Such transfers of invertebrates as well as microbiota, for example on unsterilised sampling equipment, is becoming more likely with the increasing movement of scientists and others between remotely separated parts of the region.

Human beings, and the organisms that we bring with us, are now an important new element of the biological environment in Antarctica. The new substrates that we are importing are likely to favour the associated alien microbes, or at the very least, an unusual component of the native flora. It remains to be seen whether the remarkable microbial communities that characterise this region of the world can persist in these changing habitats.

# Glossary

**actinomycetes**  A large group of branching, filamentous, usually Gram-positive bacteria that form a fungus-like mycelium. They are common in soils e.g. *Streptomyces* in moss-covered soils on Ross Island and at the thermal sites on Mount Melbourne.

**ascomycetes**  The largest class of fungi, distinguished by their sexually produced spores which are produced in a sac or acus. Common representatives include *Penicillium, Aspergillus* and the yeasts (*Endomycetales* or *Saccharomycetales*). Many yeasts, however, have lost the ability to reproduce sexually; these asporogenous yeasts include the important genera *Cryptococcus, Torulopsis* and *Candida* found in antarctic lakes and soils.

**autotrophs**  Organisms which obtain their energy from sources other than organic compounds.

**chlorophyll *a* (chl*a*)**  The green pigment that is possessed by all algae (as well as higher plants) and cyanobacteria. It mostly functions as an 'antenna pigment' absorbing PAR (see below), but a small quantity also plays a central role in the photosynthetic reaction centres. Chl*a* is often used as a measure of phototrophic biomass in natural environments.

**chlorophytes**  The green algae, characterised by chlorophylls *a* and *b*. In Antarctica this group includes snow algae (volvocales), certain lithic species (e.g. *Hemichloris*), stream algae (e.g. *Zygnema* on Signy Island) and a wide range of terrestrial species (e.g. *Prasiococcus, Prasiola*).

**chemotrophs**  Organisms which derive their energy from the oxidation of inorganic compounds (chemoautotrophs or chemolithotrophs) or organic compounds (chemoorganotrophs or heterotrophs). Chemoautotrophs include nitrifying bacteria, which form sharply defined maxima in certain antarctic lakes and which have been implicated as potentially important primary producers in the sea beneath the Ross Ice Shelf.

**choanoflagellates**   A group of marine protozoa that may play a critical role in the Southern Ocean food chain.

**chrysophyte**   The golden-brown algae, characterised by chlorophylls *a* and *c*, and often the brown carotenoid fucoxanthin. Important groups in the Antarctic include diatoms, prymnesiophytes and silicoflagellates.

**coccolithophores**   An important group of marine phytoplankton. They are members of the Prymnesiophyceae and their cells are covered by organic scales encrusted with calcite.

**cyanobacteria**   Prokaryotic phototrophs containing chlorophyll *a* and phycobilin pigments. Also known as blue-green algae or cyanophytes, these are probably the dominant life forms in the non-marine antarctic environment.

**denitrifying bacteria**   Heterotrophic anaerobes which use nitrate or nitrite as the terminal electron acceptor in their respiratory chain, reducing these compounds to nitrogen gas.

**diatoms**   Members of the algal class Bacillariophyceae, characterised by silica cell walls (frustules). Two groups are recognised: centric diatoms have a radial symmetry, and pennate diatoms have a bilateral symmetry. Pennate diatoms are especially important in the sea-ice communities, but both groups are well represented in the marine plankton and also occur in antarctic soil and freshwater habitats.

**dinoflagellates**   Biflagellated unicells of the algal division Pyrrhophyta.

**epifluorescence microscopy**   A powerful microbiological technique in which the microbial cells are stained with a fluorescent compound and examined against a black background in a fluorescence microscope. Algae and cyanobacteria can be distinguished by their autofluorescence. However the assay does not differentiate viable and non-viable cells.

**foraminifera**   A group of amoeboid protozoa, mostly marine, coated with a hard shell with a number of minute holes or one or more larger holes through which the rhizopodia protrude. These organisms are especially important in the antarctic marine benthos, but planktonic species are also well represented.

**Gram-negative and Gram-positive**   Bacteria have been traditionally divided into these two major groups on the basis of whether they retain a crystal violet-iodine complex stain. Bacteria which do not (Gram-negative) have a thin cell wall with an outer layer of lipopolysaccharide and protein. Representatives include *Streptococcus*, *Escherichia*, *Achromobacter*, *Alcaligenes* and flavobacteria which are strongly pigmented. Gram-positive cells have a much thicker wall without the outer layer, and include *Corynebacterium* and *Arthrobacter*. Some genera vary in their response to Gram-staining e.g. *Bacillus*.

**heterocyst**   A specialised cell in which nitrogen fixation is performed, possessed by some cyanobacteria (see Fig. 8.5).

**heterotrophs**   Organisms such as fungi, protozoa and animals which use organic carbon as their source of energy and carbon (chemoorganotrophs).

**holotrich**   One of three sub-classes of ciliated protozoa, characterised by uniform ciliation over the body surface; cf. spirotrichs which have a clockwise-oriented band of fused cilia (e.g. tintinnids); and cf. peritrichs which have anticlockwise oriented membranelles, and are usually attached to their substrate by a stalk (e.g. *Vorticella*). Common holotrich genera in antarctic freshwaters include *Didinium, Nassula, Lacrymaria, Spathidium, Cyclogramma, Urotricha* and *Cinetochilum*.

**methanogens**   Anaerobic bacteria which reduce compounds such as acetate or carbon dioxide in the presence of hydrogen to methane. These organisms tend to be rare in high sulphate waters (e.g. the sea, saline meromictic lakes) because of competition by sulphate reducers for acetate and hydrogen.

**methylotrophs**   Bacteria that derive their energy from the oxidation of methane. Some species co-oxidise ammonia to nitrous oxide and nitrite.

**nitrifying bacteria**   Chemoautotrophs which oxidise ammonia to nitrous oxide, nitrite and nitrate.

**Oscillatoriaceae**   The simplest of the filamentous cyanobacteria, forming unbranched trichomes (filaments) sometimes within a sheath. Members of this class such as *Phormidium, Oscillatoria* and *Lyngbya* are highly resistant to freezing and desiccation and are among the most common organisms in non-marine antarctic environments.

**phaeophytin**   A degradation product of chlorophyll that is sometimes used as a measure of detritus in marine and freshwaters.

**phagotrophs**   Organisms which derive their energy and carbon by engulfing particulate organic material, for example foraminifera. Certain phytoflagellates (algae with flagella) are also capable of this mode of nutrition.

**photosynthetically available radiation (PAR)**   Light in the waveband 400–700 nm that can be used for photosynthesis. PAR is commonly measured as micromoles of photons (=microEinsteins, $\mu$E) per unit area per time. Thus 1 $\mu$E = 6.02 × $10^{17}$ photons. Full sunlight in McMurdo Sound in December is approximately 1000 $\mu$E m$^{-2}$ s$^{-1}$.

**phototrophs**   Organisms which use light as their energy source. Those species which derive their cellular constituents from inorganic compounds are referred to as photoautotrophs (or photolithotrophs) while photoheterotrophs (photoorganotrophs) derive their carbon from organic compounds. Microalgae and/or cyanobacteria are often the dominant phototrophs in many antarctic environments.

**phycomycetes**   A subdivision of fungi that includes zoospore-producing fungi (Mastigomycotina e.g. chytrids) and the zygospore-producing fungi (Zygomycotina e.g. *Phycomyces* and *Mucor*).

**Prasinophyceae** Members of this algal class are related to the chlorophyceae and are characterised by a covering of minute scales over the cell and flagella. The most common antarctic representative *Pyramimonas* is a four-lobed cell with four flagella, and occurs in sea ice and saline lakewaters.

**Prymnesiophyceae** This algal class within the Chrysophyta contains biflagellates that often possess another appendage called a haptonema. The most common representatives are coccolithophores, and the species *Phaeocystis pouchetii* which forms blooms of floating gelatinous colonies, for example in the Ross Sea.

**psychrophile**   An organism which only grows at low temperature. The more precise criteria defined by Morita (1975) are often preferred by microbiologists: psychrophilic species grow over the range 0 °C (or less) to 20 °C (or less) with an optimum growth temperature of 15 °C or less.

**psychrotroph**   An organism which grows at low temperature but also at much higher temperatures. Temperature criteria used to separate this group vary arbitrarily from study to study. Some authors adopt the criteria: growth at 5 °C or less, but with a temperature optimum or maximum above the limits set by Morita (1975) for psychrophiles.

**radiolarians**   Rhizopod protozoa, generally marine, with stiff radiating pseudopodia and a silicon skeleton.

**testate amoebae**   This term is used in the antarctic soil literature to refer to amoeboid protozoa, classified in the rhizopod order Testacida, which have a simple, single-chamber shell. As in the temperate zone, these organisms are especially common in moss beds and peaty soils, as well as freshwater habitats. A large group of marine amoebae with a more complex test are classified in the related order Foraminiferida.

**tintinnid**   A class of marine ciliates found in the benthos and plankton. The single cells are covered by a lorica. They may be an important link in the food chain between bacteria and small-sized phytoplankton (picoplankton, cells <2 $\mu$m, and nanoplankton, <20 $\mu$m) and larger animals (e.g. krill).

**tritiated thymidine assay**  A technique to measure secondary production by bacteria. The assay measures the incorporation of $^3$H-thymidine into DNA, but the conversion to an overall estimate of production rates rests on assumptions that may not always be valid (e.g. an average bacterial genome size, presence of thymidine kinase in all bacteria).

**Volvocales**  This order within the Chlorophyceae contains motile algae often with a complex life history passing through zygote resting spores and non-motile colonial stages. Antarctic representatives include snow algae such as *Chlamydomonas* and *Chilomonas*, and a frequent constituent of hypersaline lakes, *Dunaliella*.

# Appendix 1

## Antarctic climates

The severe climate of Antarctica decisively influences the distribution and activity of the biota including the microbial life-forms of the region. This appendix first describes the general climatic features of Antarctica (pressure, wind, precipitation, humidity, radiation and temperature) and then examines the large variations in climate between different localities mentioned in the text.

### 1.1 Atmospheric pressure and average wind distribution

A low pressure trough between latitudes 60 and 70 °S continuously surrounds the continent and lies between the south subtropical anticyclones and the high pressure region centred over the polar plateau (Fig. A1.1). Winds blow from the north towards this trough and are deflected to the east by the Earth's rotation. This results in strong westerlies which average about 18–24 knots (9–12 m s$^{-1}$). The winds increase with increasing latitude towards the low pressure belt giving rise to the nautical terms 'roaring forties', 'howling fifties' and 'screeching sixties'.

A series of deep depressions moving from the northwest to the southeast are superimposed upon this general circulation. These originate between 45 and 60 °S and end in the low pressure trough where they circulate in a clockwise direction around the continent. At any one time four or five of these depressions may be detected (Fig. A1.1), typically moving at about 500 km d$^{-1}$. In the southern region of the cyclonic depressions the winds move from east to west, and this easterly flow extends towards the margins of the continent.

The anticyclonic system over the polar plateau generates a northerly wind deflected to the west by the rotation of the Earth. Wind direction over the continent, however, is mostly controlled by the local topography of ice and snow. The cold dense air flowing down the slope of the ice sheet achieves very high velocities; more than 80 knots (40 m s$^{-1}$) at the mouth

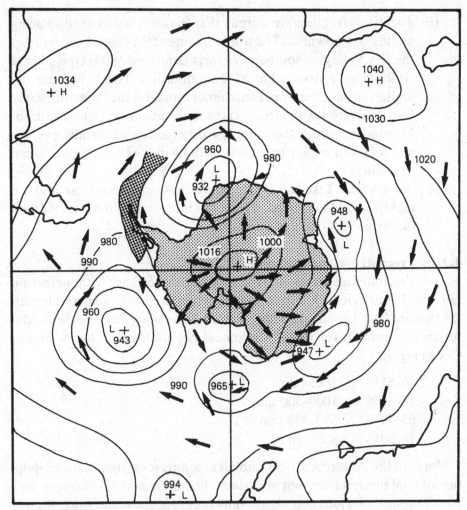

Fig. A1.1. Atmospheric pressure and wind in the antarctic region. The circumpolar distribution of low pressure systems in July 1979 (modified from Deacon, 1984) and prevailing wind directions in Antarctica (modified from Tchernia, 1980), Deacon (1984) and Walton (1984). The darker shaded area is referred to as the maritime zone.

of certain valleys. These so called katabatic winds may be experienced up to 25 km offshore. The term katabatic generally refers to any wind blowing down an incline, but three types of such winds can be distinguished (Schwerdtfeger, 1970):

(i) *Sloped inversion winds*. This is the wind type of the polar plateau so named because it results from the slope of the inversion layer. These winds blow with very small variation in speed or direction for weeks to months.

(ii) *Foehn winds*. These are warmer than the air that they are displacing as they flow downhill. The winds in some of the coastal valleys (e.g. the Dry Valleys of southern Victoria Land) are of this type and are typically very strong, but of short duration. This is because the limited air reserves upstream cannot replenish the flow sufficiently fast. Locations with strong katabatic storms like Cape Denison, Mawson and Mirny therefore also experience sudden lulls, eventually followed by abrupt onsets of the wind, with high directional constancy.

(iii) *Bora winds*. Like the Foehn winds these can be intense and are highly variable in speed. These winds are colder than the air they displace.

## 1.2    Precipitation and humidity

Precipitation across Antarctica tends to decrease with increasing latitude. It is extremely low in the centre of the continent, comparable with the Sahara and less than in the Arctic. Tchernia (1980) presents the following approximate values for annual precipitation in the Southern Ocean region:

| 40–55 °S | 1000 mm |
| 55–65 °S | 1000–500 mm |
| 65–70 °S | 500–200 mm |
| 70–90 °S | <200 mm |

North of the Antarctic Circle rainfall accounts for a significant proportion of total precipitation, but at latitudes higher than 65 °S in summer and 60 °S in winter precipitation almost only occurs as snow and frost. Rain is virtually non-existent over the antarctic continent, except at the northwestern tip of the Antarctic Peninsula. The high mountains of the peninsula deflect the westerly air masses circulating around the continent, and the western side of the peninsula and its associated islands receive greater and more frequent precipitation than other coastal sites on the continent. Blowing snow is a major component of the water budget throughout the continent.

The low temperatures of the antarctic region result in very low water vapour densities at saturation (see Appendix 2), and in the arid continental region even the relative humidity is low. These effects result in a saturation vapour pressure of surface air at Vostok of only 0.003 mb, rising to 1–4 mb at Mirny on the coast (Markov *et al.*, 1968). These values contrast with a saturation vapour pressure of about 30 mb at the equator.

### 1.3 Radiation balance and air temperature

A comparison of the radiation balance terms at 'different sites emphasises the large regional variations in the antarctic climate even on the continent (Table A1.1). Incoming global radiation (sum of direct and diffuse solar radiative fluxes) is highest on the polar plateau (see Vostok in Table A1.1), mostly due to the reduced thickness of atmosphere through which the radiation is transmitted.

Albedo (ratio of reflected radiation to global radiation) is one of the most important factors in an energy budget. As seen in Table A1.1 this varies enormously from very low values in the ice-free regions such as Novolazarezskaya and Vanda to extremely large values on snow-covered ice shelves and the continental plateau (e.g. Vostok). The extreme cold of Antarctica is largely due to the high albedo associated with the snow and ice over most of the continent (and seasonally over the ocean) which reflect about three quarters of the total incident solar energy flux.

Effective longwave radiation ($E_L$) is the sum of the upward directed (negative) terrestrial radiation and the downward directed atmospheric radiation. On the polar plateau and to a lesser extent elsewhere in the region during winter the low surface temperatures result in a relatively low emission of longwave radiation and the presence of a warmer moister layer at 500–1000 m above the surface gives rise to significant downward radiation. These small values of $E_L$ (Table A1.1) provide little energy for the heating of air above the surface and a strong temperature inversion persists (maintained by the high albedo), typically for eight months on the plateau and for the four winter months at coastal sites.

Net radiation (NR) is the sum of shortwave (mostly incoming) and longwave (mostly outgoing) radiation. NR is positive in the ice-free and negative in the region's snow and ice-covered areas (Table A1.1). Thus in the ice-free regions energy is available for snow and ice melting in summer whereas the main portion of the continent acts as a net cold sink. This continuous loss of heat is compensated by the advection of warmer maritime air masses from the north.

Mean annual air temperatures range from well above freezing on the subantarctic islands to below −50°C on the continental ice sheet (Table A1.2). Temperatures range much lower than in the Arctic. For example Markov *et al.* (1970) report extreme winter minima in Siberia of −71 °C at Oimyakon, −68 °C at Hera and −67.6 °C at Verkhoyansk. These values are at least 10 °C warmer than the extreme minima in the continental interior of Antarctica e.g. −86.8 °C at Sovetskaya, −80.7 °C at Komsomolskaya and −89.6 °C at Vostok. Mean daily air temperatures for the

Table A1.1. *Measured energy fluxes at ground level for specific sites in Antarctica*

$G$ = global radiation, $E_L$ = effective longwave radiation, and $NR$ = net radiation, each expressed as mean monthly MJ m$^{-2}$, $a$ = albedo in %, ASL = altitude above sea level.

| | Jan | Feb | Mar | Apr | May | Jun | Jul | Aug | Sep | Oct | Nov | Dec | Annual Total |
|---|---|---|---|---|---|---|---|---|---|---|---|---|---|
| *Faraday (65 °S 64 °W, 11 m ASL)* | | | | | | | | | | | | | |
| G | 552 | 349 | 185 | 73 | 20 | 4 | 12 | 67 | 187 | 407 | 556 | 654 | 3065 |
| a | 69 | 65 | 70 | 74 | 75 | 79 | 77 | 80 | 82 | 84 | 81 | 74 | — |
| $E_L$ | −89 | −86 | −78 | −68 | −79 | −65 | −76 | −72 | −67 | −76 | −79 | −106 | −941 |
| NR | 82 | 37 | −22 | −49 | −74 | −64 | −73 | −58 | −33 | −9 | 27 | 63 | −174 |
| *Novolazarezkaya (71 °S, 12 °E, 99 m ASL)* | | | | | | | | | | | | | |
| G | 832 | 485 | 248 | 69 | 4 | —[a] | — | 34 | 175 | 454 | 721 | 915 | 3937 |
| a | 20 | 21 | 25 | 26 | — | — | — | 33 | 34 | 29 | 24 | 22 | — |
| $E_L$ | −264 | −176 | −143 | −127 | −121 | −109 | −123 | −134 | −137 | −169 | −204 | −237 | −1944 |
| NR | 402 | 210 | 44 | −76 | −118 | −109 | −123 | −111 | −21 | 156 | 344 | 480 | 1078 |
| *Vanda (78 °S, 162 °E, 94 m ASL)* | | | | | | | | | | | | | |
| G | 812 | 408 | 88 | 9 | — | — | — | 4 | 39 | 298 | 701 | 800 | 3159 |
| a | 20 | 20 | 43 | 43 | — | — | — | 43 | 28 | 27 | 22 | 20 | — |
| $E_L$ | −353 | −233 | −42 | −83 | −131 | −117 | −133 | −112 | −116 | −218 | −307 | −336 | −2181 |
| NR | 297 | 93 | 8 | −78 | −131 | −117 | −133 | −110 | −88 | 0 | 240 | 304 | 285 |
| *Vostok (78 °S, 107 °E, 3488 m ASL)* | | | | | | | | | | | | | |
| G | 1087 | 601 | 224 | 18 | — | — | — | 3 | 98 | 458 | 940 | 1235 | 4664 |
| a | 83 | 84 | 85 | 86 | — | — | — | 89 | 89 | 86 | 85 | 83 | — |
| $E_L$ | −131 | −71 | −63 | −45 | −46 | −46 | −46 | −44 | −49 | −68 | −102 | −102 | −831 |
| NR | 54 | 25 | −29 | −42 | −46 | −46 | −46 | −44 | −38 | −4 | 39 | 90 | −87 |

[a] — = period of winter darkness.
Data compiled from Schwerdtfeger (1984).

Table A1.2. *Air temperature data for stations throughout the antarctic region*
The stations have been arranged in order of decreasing mean annual temperature.

| Station | Location | | | Daily air temperature | | | | | | |
| --- | --- | --- | --- | --- | --- | --- | --- | --- | --- | --- |
| | | | | | January | | | July | | |
| | | | Annual | min | mean | max | min | mean | max |
| *Subantarctic islands* | | | | | | | | | | |
| Campbell | 52° 33'S; | 169° 09'E; | 23 m | 6.8 | 7.0 | 9.4 | 11.8 | 2.4 | 4.5 | 6.6 |
| Marion | 46° 53'S; | 37° 52'E; | 23 m | 5.3 | 4.3 | 6.6 | 9.7 | 1.4 | 4.7 | 6.4 |
| Macquarie | 54° 30'S; | 158° 57'E; | 6 m | 4.7 | 5.0 | 6.7 | 8.4 | 1.2 | 3.0 | 4.5 |
| Kerguelen | 49° 21'S; | 70° 12'E; | 18 m | 4.4 | 3.7 | 7.2 | 10.8 | −0.9 | 1.8 | 4.5 |
| Heard | 53° 01'S; | 73° 23'E; | 5 m | 1.3 | 1.5 | 3.2 | 5.2 | −2.4 | −0.4 | 1.3 |
| Stanley (Falkland Islands) | 51° 45'S; | 57° 56'W; | 53 m | 5.6 | 5.2 | 8.7 | 12.7 | −0.1 | 2.1 | 4.3 |
| Grytviken (South Georgia) | 54° 17'S; | 36° 30'W; | 3 m | 1.8 | 1.5 | 4.4 | 8.0 | −4.3 | −1.5 | 1.4 |
| *Antarctic Peninsula region* | | | | | | | | | | |
| Almirante Brown (Paradise Bay, Peninsula) | 64° 53'S; | 62° 53'W; | 7 m | −2.4 | −0.9 | 1.8 | 4.3 | −9.3 | −6.8 | −4.4 |
| Bellingshausen (King George I., South Shetlands) | 62° 12'S; | 59° 0'W; | 14 m | −2.6 | −[a] | 1.1 | | | −7.1 | |
| Admiralty Bay (King George I., South Shetlands) | 62° 03'S; | 58° 24'W; | 9 m | −2.7 | −0.5 | 1.2 | 3.2 | −11.7 | −8.3 | −5.0 |
| Deception Island (South Shetlands) | 62° 59'S; | 60° 34'W; | 8 m | −3.0 | −0.6 | 1.3 | 2.8 | −11.7 | −8.6 | −5.5 |
| Signy Island (South Orkneys) | 60° 43'S; | 45° 36'W; | 7 m | −3.7 | −1.1 | 0.4 | 2.2 | −15.2 | −10.9 | −6.7 |
| Palmer Station (Anvers I. nr. Peninsula) | 64° 48'S; | 64° 06'W; | 8 m | −3.9 | | 1.4 | | | −10.8 | |

*continued*

Table A1.2 *continued*

| Station | Location | Annual | January | | | July | | |
|---|---|---|---|---|---|---|---|---|
| | | | min | mean | max | min | mean | max |
| General Bernardo O'Higgins (Peninsula) | 63° 18'S; 57° 54'W; 10 m | −4.0 | — | 0.5 | — | — | −9.2 | — |
| Faraday (Argentine I. nr. Peninsula) | 65° 15'S; 64° 15'W; 11 m | −5.4 | −1.8 | 0.1 | 2.1 | −16.2 | −11.8 | −8.2 |
| Esperanza (Peninsula) | 63° 24'S; 56° 59'W; 7 m | −5.8 | −2.4 | 0.4 | 3.0 | −15.0 | −9.9 | −6.3 |
| Rothera (Adelaide I. nr. Peninsula) | 67° 36'S; 68° 06'W; 15 m | −6.4 | — | 0.7 | — | — | −15.3 | — |
| Marguerite Bay (Peninsula) | 68° 11'S; 67° 01'W; 9 m | −7.3 | −2.5 | 0.1 | 2.6 | −16.3 | −11.7 | −7.7 |
| *Continental margin* | | | | | | | | |
| Oasis Station (Bunger Hills) | 66° 18'S; 100° 43'E; 28 m | −9.1 | — | 1.8 | — | — | −16.8 | — |
| Casey (East Antarctica) | 66° 20'S; 110° 30'E; 12 m | −9.2 | — | 0.1 | — | — | −14.6 | — |
| Davis (Vestfold Hills) | 68° 35'S; 77° 58'E; 12 m | −10.3 | −2.2 | −0.2 | 1.6 | −20.4 | −17.4 | −14.7 |
| Syowa (Coastal Island) | 69° 00'S; 39° 35'E; 15 m | −10.5 | −4.4 | −1.2 | 1.2 | −22.1 | −18.3 | −14.7 |
| Novolazarezkaya | 70° 46'S; 11° 50'E; 87 m | −10.6 | — | −0.7 | — | — | −17.7 | — |
| Dumont d'Urville I. (5 km from continent) | 66° 40'S; 140° 01'E; 41 m | −10.7 | −3.1 | −0.5 | 1.5 | −21.0 | −17.7 | −14.4 |
| Molodezhnaya | 67° 40'S; 45° 51'E; 42 m | −10.9 | — | −0.5 | — | — | −18.0 | — |
| Mirny (rock outcrop) | 66° 33'S; 93° 01'E; 35 m | −11.3 | — | −1.5 | — | — | −16.5 | — |
| Mawson (ice-free rock outcrop) | 67° 36'S; 62° 53'E; 8 m | −11.3 | −3.1 | −0.3 | 2.2 | −20.4 | −17.8 | −15.1 |
| Cape Denison | 67° 06'S; 142° 42'E; 6 m | −12.8 | — | −0.9 | — | — | −19.6 | — |
| Leningradska | 69° 30'S; 159° 24'E; 295 m | −14.4 | — | −4.0 | — | — | −20.8 | — |
| Roi Baudouin | 70° 26'S; 24° 19'E; 37 m | −15.2 | — | −4.5 | — | — | −23.1 | — |
| Hallett (Ross Sea region) | 72° 18'S; 170° 18'E; 5 m | −15.3 | −3.3 | −1.1 | 1.1 | −29.4 | −26.7 | −23.9 |

| Station | Coordinates | Elevation | | | | | | | |
|---|---|---|---|---|---|---|---|---|---|
| Marble Point (AWS)[b] (Scott Coast) | 77° 24'S; 163° 48'E; | 121 m | −16.8 | −12.4 | −3.4 | 5.4 | −41.6 | −25.4 | −6.9 |
| Sanae | 70° 20'S; 02° 25'W; | 52 m | −17.1 | −10.3 | −4.1 | −1.4 | −33.3 | −27.1 | −22.6 |
| McMurdo Station (Ross Island, nr. continent) | 77° 51'S; 166° 37'E; | 24 m | −17.4 | −6.1 | −3.2 | −1.1 | −31.7 | −26.8 | −21.7 |
| Halley Bay (Brunt Ice Shelf, Weddell Sea) | 75° 31'S; 26° 38'W; | 32 m | −18.5 | −8.7 | −5.3 | −2.7 | −31.7 | −27.7 | −23.7 |
| Vanda (Dry Valleys of s. Victoria Land) | 77° 30'S; 161° 36'E; | 94 m | −19.8 | −2.2 | 1.2 | 4.6 | −42.6 | −38.0 | −33.4 |
| Scott Base (Ross I. 30 km from continent) | 77° 51'S; 166° 45'E; | 16 m | −20.2 | −8.5 | −5.0 | −1.4 | −35.5 | −29.3 | −23.0 |
| Ellsworth (ice shelf on Weddell Sea) | 77° 44'S; 41° 08'W; | 43 m | −22.9 | −14.6 | −7.1 | −1.7 | −37.1 | −32.9 | −28.5 |
| Little America V | 78° 11'S; 162° 12'W; | 42 m | −23.8 | −9.4 | −6.7 | −4.4 | −38.9 | −34.4 | −30.0 |
| Siple (Ellsworth Land) | 75° 54'S; 84° 12'W; | 1050 m | −24.8 | — | −11.9 | — | — | −30.5 | −30.5 |
| Meeley (AWS)[b] (Northwest Ross Ice Shelf) | 78° 30'S; 170° 12'E; | 49 m | −24.9 | −18.4 | −7.5 | 2.7 | −55.4 | −36.2 | −16.6 |
| *Continental interior* | | | | | | | | | |
| Byrd (Marie Byrd Land) | 79° 59'S; 120° 00'W; | 1528 m | −27.9 | −18.8 | −13.1 | −15.8 | −41.8 | −37.5 | −32.9 |
| Mizuho | 70° 42'S; 44° 18'E; | 2230 m | −32.3 | −18.6 | — | — | — | −39.1 | −39.1 |
| Amundsen–Scott station (South Pole) | 90° 00'S; | 2835 m | −49.3 | −29.4 | −28.3 | −26.7 | −61.7 | −58.3 | −55.6 |
| Dome C (AWS)[b] (East Antarctica) | 74° 30'S; 123° 00'E; | 3280 m | −50.5 | — | −30.0 | — | — | — | −58.5 |
| Vostok II | 78° 28'S; 106° 48'E; | 3488 m | −55.4 | — | −32.7 | — | — | — | −67.0 |
| Plateau Station | 79° 12'S; 40° 30'E; | 3625 m | −56.4 | — | −33.9 | — | — | — | −68.0 |

[a] — = data not available.

[b] AWS = Automatic Weather Station.

Data are from Phillpot (1967), Solopov (1969), New Zealand Meteorological Service (1983), Schwerdtfeger (1984) and Savage & Stearns (1985).

warmest month (January) and coldest month (July) at the South Pole lie well below the values for the equivalent months at the North Pole: $-28$ and $-58\,°C$ at $90\,°S$, but 0 and $-36\,°C$ at $90\,°N$ (Markov *et al.*, 1970). The high altitude of the thick ice sheet forming the antarctic continental plateau, and the relatively warm Arctic Ocean in the north polar region are the major factors controlling these marked differences.

## 1.4    Climatic zones

A wide range of classification schemes have been proposed to separate different climatic zones of the antarctic region. The most useful classification for microbiological discussions is that described by Lewis Smith (1984) who recognises three geobotanical regions from their biotic as well as climatic characteristics:

(i) The subantarctic region experiences a wet, cool oceanic climate with mean temperatures above freezing for at least six months and precipitation over 900 mm $y^{-1}$. The vegetation is tundra-like and there are spiders, molluscs, earthworms, many insects and some land birds. Representative localities include the first seven sites listed in Table A1.2.

(ii) The maritime antarctic region (Fig. A1.1) is characterised by mean daily summer temperatures above $0\,°C$ and winter means rarely below $-10$ to $-15°C$. The vegetation is dominated by non-vascular plants but two angiosperms, the grass *Deschampsia antarctica* and the small herb *Colobanthus quitensis* form closed swards as far south as Marguerite Bay ($68\,°S$). The maritime region may be subdivided into the northern province which includes the South Sandwich, South Orkney and South Shetland Islands, the west coast of the Antarctic peninsula to $66\,°S$, and Bouvetøya; and the southern province which includes the west coast of the Antarctic Peninsula and associated islands from $66$–$70\,°S$, and the northeast coast of the peninsula to *c.* $63\,°S$. The northern province is characterised by 350–500 mm of precipitation each year with much of it as rain in summer while the southern province is drier ($<350$ mm), cooler and contains a much less diverse biota.

(iii) The continental antarctic region is cold and arid and microbial life forms (including microalgae) generally dominate the biota. The region can be separated into three provinces characterised by their temperature and wind regimes:

(a) *Coastal.* Summer temperatures exceed freezing for up to one month and winter means range from $-5$ to $<-20\,°C$. The semi-desert vegetation is restricted in species composition and coverage and contains mosses and lichens. This province includes the coastal fringes of the continent, the east coast of the peninsula, the ice-free areas such as the Dry Valleys' of southern Victoria Land, and offshore islands such as Balleny and Scott Islands. All of these regions experience strong katabatics as well as oceanic influences, but to a variable extent.

(b) *Slope*. All mean monthly temperatures are below −5 °C and precipitation is around 100 mm of water equivalent. Katabatics are also a feature of this category. Occasional patches of lichens and mosses are the only vegetation. This category incorporates the mountain and glacier zones inland from the coast including isolated nunataks in the plateau.

(c) *Ice plateau*. Mean monthly temperatures are below −15 °C, falling below −30 °C in winter. Precipitation is minimal and the altitude is typically greater than 2000 m. Here the only life forms are microbial and even these organisms are rarely likely to be active. This region experiences the lighter and more constant sloped inversion winds.

## 1.5    Regional climate summaries

This section examines the range of climates through the antarctic region by way of six illustrative localities.

### 1.5.1    South Orkney Islands

The South Orkneys lie within the northern maritime antarctic region. The islands are humid and windy (mean wind speed of 6 m s$^{-1}$ at Signy) and frequently experience precipitation, fog and cloud cover. Radiation input is therefore low in all seasons. Precipitation occurs on about 335 days of the year with an annual total of about 400 mm (Allen *et al*. 1967). The annual mean temperature at Signy Island Station (United Kingdom) is warm (−3.5 °C) by comparison with the continental sites. January, February and March, and sometimes also December and April have mean daily temperatures above 0 °C (Table A1.3).

### 1.5.2    Paradise Bay, Antarctic Peninsula

The western side of the Antarctic Peninsula also experiences a maritime climate with rain in summer and relatively warm temperatures (Table A1.4). The mean annual air temperature at Almirante Brown Station (Argentina), as well as temperatures in the coldest month (July)

Table A1.3. *Climatic data for Signy Island station in the South Orkneys*

| | Jan | Feb | Mar | Apr | May | Jun | Jul | Aug | Sep | Oct | Nov | Dec |
|---|---|---|---|---|---|---|---|---|---|---|---|---|
| *Air temperature* (°C)[a] | | | | | | | | | | | | |
| mean | 0.4 | 0.7 | 0.6 | −2.3 | −6.9 | −8.5 | −10.9 | −9.1 | −5.1 | −2.1 | −1.8 | −0.4 |
| mean daily maximum | 2.2 | 2.7 | 2.2 | 0.1 | −3.3 | −4.7 | −6.7 | −5.6 | −2.0 | −0.1 | 0.7 | 1.3 |
| mean daily minimum | −1.1 | −1.0 | −1.8 | −4.8 | −10.1 | −12.6 | −15.2 | −13.1 | −8.7 | −4.8 | −3.2 | −1.9 |
| *Cloud cover* (0 = clear, 8 = overcast) | 7.2 | 7.2 | 7.3 | 7.0 | 6.5 | 6.1 | 5.7 | 6.0 | 6.5 | 6.9 | 7.3 | 7.4 |
| *Precipitation*[b] (mm water) | 19 | 24 | 47 | 25 | 9 | 5 | 11 | 26 | 3 | 5 | 21 | 44 |
| *Global radiation*[c] (MJ m$^{-2}$ d$^{-1}$) | 14.1 | 10.3 | 4.5 | 2.6 | 1.0 | 0.5 | 0.8 | 2.4 | 7.5 | 11.7 | 18.1 | 17.0 |

[a] Data are means for 1947–60 (Phillpot, 1967).
[b] Data are for 1979 (Davis, 1986).
[c] Walton (1977).

Table A1.4. Climatic data for Almirante Brown station (Argentina) at Paradise Bay on the Antarctic Peninsula

| | Jan | Feb | Mar | Apr | May | Jun | Jul | Aug | Sep | Oct | Nov | Dec |
|---|---|---|---|---|---|---|---|---|---|---|---|---|
| *Air temperature* (°C)[a] | | | | | | | | | | | | |
| mean | 1.8 | 1.1 | -0.4 | -2.0 | -4.2 | -5.8 | -6.8 | -5.7 | -4.8 | -2.0 | -0.6 | 0.7 |
| mean daily maximum | 4.3 | 3.7 | 3.6 | 1.0 | -2.0 | -3.0 | -4.4 | -3.3 | -2.6 | 0.3 | 2.1 | 2.9 |
| mean daily minimum | -0.9 | -0.6 | -1.8 | -2.8 | -6.2 | -7.7 | -9.3 | -8.3 | -8.3 | -4.4 | -2.9 | -1.3 |
| Cloud cover (0 = clear, 8 = overcast)[a] | 6.5 | 6.8 | 6.4 | 6.4 | 5.8 | 6.3 | 5.8 | 6.4 | 6.3 | 6.9 | 6.7 | 6.1 |

[a]Data are for 1951–56 (compiled from Phillpot, 1967).

and warmest month (Janury) lie above those experienced at Signy Island (Table A1.4) despite its location about 450 km to the south of Signy Island. In general this western side is several degrees warmer and much less windy than the eastern side of the peninsula.

### 1.5.3    *McMurdo Sound region*

Continuous meteorological records from two coastal stations on Ross Island illustrate the wide variations that can be imposed by local topography and exposure (Table A1.5). Only two kilometres separate Scott Base (New Zealand) from McMurdo Station (USA), but the wind regime, temperature, and incidence of snow and fog differ significantly between the two sites. Scott Base is more exposed to cold winds from the south and its mean annual air temperature ($-20\,°C$) is about 3 °C cooler than at McMurdo Station. Local topography restricts 60% of the observed winds at Scott Base to a narrow sector between 010° and 040° whereas at McMurdo Station the winds come from a much broader easterly sector. Strong winds are typically from the south; 85% of all winds over 30 knots at Scott Base are from this direction. Unlike McMurdo Station, Scott Base also experiences very strong northwesterlies. Snow occurs on an average of 72 d $y^{-1}$ at Scott Base and 62 d $y^{-1}$ at McMurdo Station. Fog is also more frequent at Scott Base (16 d $y^{-1}$) than at McMurdo Station (9 d $y^{-1}$) (New Zealand Meteorological Service, 1983).

On the other side of McMurdo Sound at Marble Point the mean annual air temperature ($-17\,°C$) is comparable with McMurdo Station. Winter daily minimum temperatures average about 6 °C colder than Scott Base and daily summer maxima average about 6 °C warmer. Sustained periods of thaw are much more likely at this continental site. The highest mean daily maximum at Scott Base is $-1.4\,°C$ in January (Table A1.5), but at Marble Point the mean daily maximum is above 0 °C in January, February, April and December, and lies less than 1 °C below freezing in March, June, October and November (Savage & Stearns, 1985). Further to the south, the air in the boundary layer over the Ross Ice Shelf averages 5–10 °C cooler temperatures than at McMurdo Station (see Meeley Station in Table A1.2).

### 1.5.4    *'Dry Valleys' region of southern Victoria Land*

The so-called 'Dry Valleys' (which more recently have gained another as yet unofficial name, the 'Ross Desert') lie on the edge of the continent adjacent to McMurdo Sound and encompass an area of about 5000 km². The tall mountains (>2000 m) and deeply cut valleys of this

Table A1.5. Climatic data from Scott Base and McMurdo Station in the McMurdo Sound region

| | Jan | Feb | Mar | Apr | May | Jun | Jul | Aug | Sep | Oct | Nov | Dec |
|---|---|---|---|---|---|---|---|---|---|---|---|---|
| *Air temperature (°C)*[a] | | | | | | | | | | | | |
| mean | −5.1 | −10.7 | −20.4 | −24.1 | −27.2 | −26.3 | −29.3 | −30.8 | −28.1 | −22.7 | −11.9 | −5.3 |
| mean daily maximum | −1.4 | −7.2 | −15.8 | −18.8 | −21.0 | −20.2 | −23.0 | −24.1 | −22.0 | −17.0 | −7.5 | −1.9 |
| mean daily minimum | −8.5 | −14.2 | −24.9 | −29.4 | −33.4 | −32.4 | −35.5 | −37.4 | −34.2 | −28.3 | −16.3 | −8.8 |
| *Cloud cover (0 = clear, 8 = overcast)*[b] | 4.1 | 5.3 | 5.5 | 4.7 | 3.8 | 4.0 | 3.5 | 3.2 | 4.6 | 4.5 | 3.9 | 4.4 |
| *Snow (days per month)*[a] | 7 | 9 | 7 | 6 | 6 | 5 | 5 | 5 | 5 | 6 | 7 | 8 |
| *Fog (days per month)*[a] | 2 | 1 | 1 | 2 | 1 | 1 | 2 | 2 | 3 | 1 | <1 | 1 |
| *Gales (days per month)*[a] | <1 | 1 | 2 | 2 | 3 | 3 | 3 | 3 | 3 | 2 | 1 | 1 |
| *Global radiation* (MJ $m^{-2}$ $d^{-1}$)[a] | 26.9 | 14.1 | 4.9 | 0.5 | 0 | 0 | 0 | 0.1 | 2.8 | 12.3 | 24.7 | 30.9 |
| *Mean wind speed* (m $s^{-1}$)[c] | 5.3 | 7.0 | 7.3 | 6.1 | 6.9 | 7.2 | 6.5 | 6.4 | 6.9 | 6.2 | 5.4 | 6.5 |

[a] Data are for 1957–80 at Scott Base (New Zealand Meteorological Service, 1983).
[b] Data are for 1956–61, at McMurdo Station (Phillpot, 1967).
[c] Data are for McMurdo Station (Schwerdtfeger, 1970).

region receive very low precipitation and apart from alpine glaciers and frozen lakes the area is largely devoid of snow and ice. Warm temperatures (up to 15 °C) and persistent up and down valley winds help keep the valleys substantially free of snow in summer. The winters are characterised by long periods of cold calm weather interspersed with periods of warm, westerly katabatic gales. Relative humidity often drops below 10% during these westerlies which can rise to more than 100 knots (Bromley, 1985). There are large microclimatic differences in summer between the individual valleys, but the only manned weather station during winter has been Vanda Station (New Zealand) in the Wright Valley (Table A1.6). The mean annual temperature at this site (−20 °C) is comparable with the coastal stations on Ross Island (see McMurdo Sound region), but Vanda summer temperatures are about 7 °C warmer and about the same amount cooler in winter. The coldest winter temperatures on record at Vanda are −56.9 °C in July and −55.2 °C in August (Bromley, 1985). Snowfalls are infrequent events that vary enormously between years in magnitude and duration (Table A1.6).

### 1.5.5    Vestfold Hills

The Vestfold Hills is an ice-free area of approximately 500 km$^2$ on the coast of Princess Elizabeth Land in East Antarctica. The region experiences a moderate snowfall, and may be completely snow-covered by March. The katabatic wind dissipates over the 10–20 km between the continental ice-sheet and the coast, and mean wind speeds at the coastal Davis Station (Australia) are relatively slow (Table A1.7). The strongest winds are from the north and originate in the major depressions which form over the South Atlantic and southern Indian Oceans, frequently in the vicinity of Marion Island (47 °S, 37 °E). Strong winds (>17 m s$^{-1}$) occur less than 4% of the time, and are most frequent in winter. The incidence of calms is relatively high (10%) by comparison with many antarctic stations.

### 1.5.6    Polar plateau

The climate of the polar plateau is extremely cold but the skies are often clear and the region experiences high levels of insolation in summer (G in Table A1.1). The coldest air temperatures on Earth have been recorded at Vostok Station (USSR) in the middle of the polar plateau (Table A1.8); the record extreme is currently −89.6 °C in July 1983 (Keys, 1984). The mean air temperatures at Vostok (−55.3 °C for 1958–9), lie well below those recorded at Amundsen–Scott Station at the South Pole

Table A1.6. *Climatic data from Vanda Station in the Dry Valley region of southern Victoria Land*

| | Jan | Feb | Mar | Apr | May | Jun | Jul | Aug | Sep | Oct | Nov | Dec |
|---|---|---|---|---|---|---|---|---|---|---|---|---|
| *Temperature* (°C)[a] | | | | | | | | | | | | |
| mean | 1.2 | −5.9 | −20.4 | −29.7 | −29.2 | −30.0 | −38.0 | −32.3 | −31.2 | −15.7 | −6.5 | 0.3 |
| mean daily maximum | 4.6 | −2.1 | −15.6 | −25.4 | −23.5 | −25.1 | −33.4 | −26.5 | −25.7 | −10.0 | −1.9 | 3.7 |
| mean daily minimum | −2.2 | −9.7 | −25.1 | −34.0 | −34.8 | −34.8 | −42.6 | −38.1 | −36.7 | −21.4 | −11.1 | −3.1 |
| *Mean windspeed* (m s$^{-1}$)[b] | | | | | | | | | | | | |
| 1969 | 6.4 | 6.2 | 3.9 | 1.5 | 1.9 | 2.1 | 4.6 | 2.3 | 3.2 | 4.9 | 7.2 | 6.6 |
| 1970 | 6.5 | 6.1 | 3.4 | 1.7 | 7.3 | 5.1 | 1.5 | 5.0 | 5.1 | 7.7 | 6.4 | 7.2 |
| *Cloud cover* (0 = clear, 8 = overcast)[b] | | | | | | | | | | | | |
| 1969 | 3.8 | 3.6 | 4.2 | 2.9 | 4.0 | 3.2 | 2.2 | 3.3 | 3.8 | 3.5 | 4.3 | 5.6 |
| 1970 | 4.3 | 4.6 | 2.5 | 3.6 | 2.2 | 2.1 | 1.3 | 2.5 | 3.5 | 3.3 | 2.8 | 4.0 |
| *Days of snowfall*[b] | | | | | | | | | | | | |
| 1969 | 3 | 4 | 8 | 4 | 9 | 7 | 3 | 6 | 0 | 2 | 2 | 11 |
| 1970 | 1 | 3 | 0 | 4 | 0 | 0 | 0 | 0 | 0 | 0 | 0 | 3 |
| *Total snowfall* (mm)[c] | | | | | | | | | | | | |
| 1969 | T[d] | T | 13 | 5 | 16 | 16 | T | 16 | 0 | 3 | T | 13 |
| 1970 | T | 2 | 0 | 5 | 0 | 0 | 0 | 0 | 0 | 0 | 0 | T |
| *Relative humidity* (%)[b] | | | | | | | | | | | | |
| 1969 | 46 | 53 | 61 | 71 | 80 | 78 | 71 | 83 | 77 | 65 | 39 | 38 |
| 1970 | 30 | 40 | 38 | 63 | 42 | 49 | 63 | 54 | 54 | 31 | 28 | 32 |

[a]Data are for 1968–80 (Nov–Jan) and 1969, 1970 and 1974 (Feb–Oct) (New Zealand Meteorological Service, 1984).
[b]Data are from Thompson, Craig & Bromley (1971).
[c]Data are from Bromley (1985).
[d]T = trace.

Table A1.7. *Climatic data for Davis in the Vestfold Hills*

| | Jan | Feb | Mar | Apr | May | Jun | Jul | Aug | Sep | Oct | Nov | Dec |
|---|---|---|---|---|---|---|---|---|---|---|---|---|
| *Air temperature* (°C)[a] | | | | | | | | | | | | |
| mean | −0.2 | −2.8 | −8.1 | −12.3 | −15.0 | −15.8 | −17.4 | −17.0 | −16.9 | −11.9 | −5.4 | −0.6 |
| mean daily maximum | 1.6 | −0.8 | −5.8 | −10.0 | −12.3 | −13.0 | −14.7 | −14.2 | −13.7 | −9.4 | −3.3 | 1.4 |
| mean daily minimum | −2.2 | −5.1 | −10.9 | −14.7 | −18.0 | −18.7 | −20.4 | −20.2 | −19.9 | −15.3 | −8.1 | −2.6 |
| *Wind speed* (m s$^{-1}$)[b] | | | | | | | | | | | | |
| mean | 4.9 | 5.3 | 5.2 | 4.4 | 4.6 | 5.1 | 5.0 | 5.1 | 4.7 | 5.0 | 5.7 | 5.5 |
| maximum | 24.8 | 30.2 | 31.1 | 36.5 | 33.9 | 37.2 | 37.2 | 37.4 | 37.9 | 37.2 | 33.6 | 31.8 |
| *Cloud cover* (0 = clear, 8 = overcast)[a] | 5.3 | 5.4 | 5.8 | 5.3 | 5.0 | 4.7 | 5.4 | 5.2 | 4.9 | 5.2 | 5.0 | 5.3 |
| *Relative humidity* (%)[b] | 60 | 57 | 62 | 67 | 66 | 66 | 69 | 68 | 65 | 65 | 60 | 60 |
| *Global radiation* (MJ m$^{-2}$ d$^{-1}$)[c] | 24.8 | 16.0 | 7.8 | 2.2 | 0.4 | 0.0 | 0.2 | 1.6 | 6.8 | 13.8 | 23.2 | 26.4 |

[a]Data are for 1957–63 (Phillpot, 1967).
[b]Data are means for 1957–64 and 1967–75 (Burton & Campbell, 1980).
[c]Data are for 1977 (Burton & Campbell, 1980).

Table A1.8. Climatic data from Vostok station on the Polar Plateau

| | Jan | Feb | Mar | Apr | May | Jun | Jul | Aug | Sep | Oct | Nov | Dec |
|---|---|---|---|---|---|---|---|---|---|---|---|---|
| *Air temperature* (°C)[a] | | | | | | | | | | | | |
| mean | −32.7 | −44.2 | −54.9 | −62.8 | −63.6 | −67.0 | −66.9 | −70.6 | −67.3 | −58.1 | −43.9 | −32.2 |
| extreme maximum | −21.1 | — | — | — | — | −53.7 | −50.1 | −52.1 | — | −42.2 | −33.5 | −21.0 |
| extreme minimum | −47.9 | −61.9 | −67.4 | −73.0 | −77.9 | −80.7 | −82.6 | −87.4 | −82.3 | −72.7 | −59.0 | −45.5 |
| *Cloud cover* (0 = clear, 8 = overcast)[a] | 3.8 | 3.0 | 4.4 | 2.9 | 4.2 | 2.5 | 2.3 | 3.0 | 3.9 | 3.2 | 3.4 | 2.6 |
| *Relative humidity* (%)[b] | 75 | 72 | 71 | 68 | 68 | 68 | 69 | 69 | 69 | 71 | 72 | 74 |
| *Mean wind speed* (m s[−1])[b] | 4.4 | 4.5 | 5.4 | 5.3 | 5.4 | 5.3 | 5.3 | 5.2 | 5.5 | 5.3 | 5.0 | 4.7 |
| *Days of drifting snow*[b] | 1 | 1 | 5 | 4 | 6 | 8 | 7 | 7 | 6 | 5 | 2 | 1 |

[a]Data are for 1958–59 (Phillpot, 1967).
[b]Data are for 1958–68 (Schwerdtfeger, 1970).

($-49.4\,°C$ for 1957–60, Phillpot, 1967), which is at slightly lower elevation. The south polar 'summer' is extremely brief with cooling beginning very rapidly in late January, a factor that contributed to the tragic end of Scott's final expedition. Low temperatures then persist throughout the rest of the year; at Vostok (Table A1.8) mean daily air temperatures remain below $-50\,°C$ for eight months. This continuous season of extreme cold is a feature of most of the continental antarctic region and is commonly referred to as the 'coreless winter'. Wind speeds on the high plateau are slow by comparison with stations on the slope and at the coast subject to strong katabatics. For example, the annual mean wind speed is $5.1$ m s$^{-1}$ at Vostok, but $10.3$ m s$^{-1}$ at Mizuho on the slope and $19.5$ m s$^{-1}$ at Cape Denison on the coast (Schwerdtfeger, 1984).

# Appendix 2

## General environmental data

Table A2.1. *Morphometric data for Antarctica*

The estimates have been derived primarily from airborne radio echo-sounding measurements.

| Coastline length | km | % |
| --- | --- | --- |
| Ice shelf | 14 110 | 44 |
| Ice stream/outlet glacier | 3 954 | 13 |
| Ice walls | 12 156 | 38 |
| Rock | 1 656 | 5 |
| Total | 31 876 | — |

| Areas | $10^3$ km$^2$ | % |
| --- | --- | --- |
| Antarctic Peninsula | | |
|   Ice | 447 | 80 |
|   Rock | 75 | 20 |
|   Total | 522 | — |
| East Antarctica | | |
|   Ice | 10 153 | 98 |
|   Rock | 201 | 2 |
|   Total | 10 354 | — |
| West Antarctica | | |
|   Ice | 1 918 | 97 |
|   Rock | 56 | 3 |
|   Total | 1 974 | — |
| Conterminus Antarctica[a] | | |
|   Ice | 13 586 | 98 |
|   Rock (ice-free) | 332 | 2 |
|   Total | 13 918 | — |

| Volumes | $10^3$ km$^3$ | % |
| --- | --- | --- |
| East Antarctica | 26 039 | 86 |
| West Antarctica | 3 262 | 11 |
| Antarctic Peninsula | 227 | 1 |
| Conterminus Antarctica | | |
|   Grounded ice | 29 378 | 98 |
|   Ice shelves | 732 | 2 |
|   Total | 30 110 | — |

[a] Includes offshore islands joined by sea ice.
From Drewry, Jordan & Jankowski (1982).

Table A2.2. *Antarctic ice shelves with areas of 10 000 km² or greater*

| Ice Shelf | Area ($10^3$ km²) | Frontal length (km) | Accumulation rate (g cm$^{-2}$ y$^{-1}$) | Ice movement (m y$^{-1}$) |
|---|---|---|---|---|
| Ross | 525 | 920 | 17.5 | 1240 |
| Filchner | 433 | 1030 | 18 | 1260 |
| Larsen | 63 | 600 | — | — |
| Fimbulisen–Bellingshausen | 40 | 760 | 40 | — |
| Amery | 39 | 250 | 28 | 800 |
| Shackleton | 37 | 740 | 70 | 700 |
| Riiser–Larsen | 32 | 940 | — | — |
| Baudouin | 31 | 1200 | 42 | 300 |
| West | 30 | 760 | 60 | 600 |
| Eights | 30 | 460 | — | — |
| Getz | 22 | 250 | — | — |
| Peacock | 20 | 160 | — | — |
| George VI | 18 | 140 | — | — |
| Wilkins | 16 | 190 | — | — |
| Exasperation | 15 | 180 | — | — |
| Thwaites | 13 | 380 | — | — |
| Salsberger | 12 | 130 | — | — |
| Elbertisen | 11 | 105 | — | — |
| Ekstremisen | 10 | 270 | — | — |

From Barkov (1985).

Table A2.3. *Physical properties of various substances at 0 °C*

| Material | Density (g cm$^{-3}$) | Specific heat (cal g$^{-1}$ °C$^{-1}$) | Heat capacity (cal cm$^{-3}$ °C$^{-1}$) | Thermal diffusivity (cm² s$^{-1}$) |
|---|---|---|---|---|
| air | $1.29 \times 10^{-3}$ | 0.24 | $3 \times 10^{-4}$ | $1.9 \times 10^{-1}$ |
| freshwater | 1.00 | 1.01 | 1.01 | $1.3 \times 10^{-3}$ |
| freshwater ice | 0.92 | 0.50 | 0.46 | $1.1 \times 10^{-2}$ |
| seawater | 1.03 | 0.93 | 0.96 | $1.4 \times 10^{-3}$ |
| snow (uncompacted) | 0.10 | 0.50 | 0.05 | $3.6 \times 10^{-3}$ |
| snow (compacted) | 0.90 | 0.50 | 0.45 | $1.2 \times 10^{-2}$ |
| granite | 2.7 | 0.19 | 0.51 | $1.3 \times 10^{-2}$ |
| sandstone[a] | 2.6 | 0.21 | 0.55 | $1.1 \times 10^{-2}$ |
| marble[a] | 2.7 | 0.21 | 0.57 | $1.0 \times 10^{-2}$ |
| dry quartz sand[a] | 1.8 | 0.19 | 0.34 | $2.0 \times 10^{-2}$ |

[a] at 20 °C.
Compiled from List (1951).

Table A2.4. *Physical properties of ice and water*

| Temperature (°C) | Specific heat of water (cal g$^{-1}$°K$^{-1}$) | Specific heat of ice (cal g$^{-1}$°K$^{-1}$) | Latent heat of fusion of water (cal g$^{-1}$) | Latent heat of evaporation of water (cal g$^{-1}$) |
|---|---|---|---|---|
| 30 | 0.998 | — | — | 580 |
| 20 | 0.999 | — | — | 586 |
| 10 | 1.001 | — | — | 592 |
| 0 | 1.007 | 0.503 | 79.7 | 597 |
| −10 | 1.02 | 0.485 | 74.5 | 603 |
| −20 | 1.04 | 0.468 | 69.0 | 609 |
| −30 | 1.08 | 0.450 | 63.0 | 615 |
| −40 | 1.14 | 0.433 | 56.3 | 622 |
| −50 | 1.3 | 0.415 | 48.6 | 629 |
| −60 | — | 0.397 | — | — |
| −70 | — | 0.380 | — | — |
| −80 | — | 0.363 | — | — |
| −90 | — | 0.346 | — | — |

Compiled from List (1951). 1 cal = 4.186 J.

Table A2.5. *Density of air at 1000 mb total pressure ($\varrho$), saturation vapour pressure ($V$) and water vapour density at saturation ($w$) as a function of temperature*

| Temperature (°C) | $\varrho$ (kg m$^{-3}$) | Over water V (mb) | Over water w (g m$^{-3}$) | Over ice V (mb) | Over ice w (g m$^{-3}$) |
|---|---|---|---|---|---|
| 40 | 1.113 | 73.777 | 51.19 | — | — |
| 30 | 1.149 | 42.430 | 30.38 | — | — |
| 20 | 1.188 | 23.373 | 17.30 | — | — |
| 10 | 1.230 | 12.272 | 9.399 | — | — |
| 0 | 1.275 | 6.108 | 4.847 | 6.107 | 4.847 |
| −10 | 1.324 | 2.863 | 2.358 | 2.597 | 2.139 |
| −20 | 1.376 | 1.254 | 1.074 | 1.032 | 0.8835 |
| −30 | 1.433 | 0.509 | 0.4534 | 0.3798 | 0.3385 |
| −40 | 1.494 | 0.189 | 0.1757 | 0.1283 | 0.1192 |
| −50 | 1.561 | 0.063 | 0.0617 | 0.0394 | 0.0382 |
| −60 | 1.634 | — | — | 0.0108 | 0.0110 |
| −70 | 1.715 | — | — | 0.0026 | 0.0028 |
| −80 | 1.804 | — | — | 0.0005 | 0.0006 |
| −90 | 1.902 | — | — | 0.0001 | 0.0001 |

Derived from List (1951).

Table A2.6. *Significant dates for the duration of daylight at various latitudes*

Civil twilight is when the sun is less than 6° below the horizon.

|  | 70 °S | 75 °S | 80 °S | 85 °S | 90 °S |
|---|---|---|---|---|---|
| Civil twilight begins | — | Jul 20 | Aug 10 | Aug 26 | Sep 8 |
| Sun returns | Jul 19 | Aug 9 | Aug 25 | Sep 8 | Sep 21 |
| Sun stays above horizon from: | Nov 19 | Nov 2 | Oct 18 | Oct 5 | Sep 21 |
| Sun stays above horizon until: | Jan 26 | Feb 12 | Feb 25 | Mar 10 | Mar 23 |
| Sun remains below horizon from: | May 25 | May 5 | Apr 19 | Apr 4 | Mar 23 |
| No civil twilight beyond: | — | May 27 | May 5 | Apr 20 | Apr 6 |

Compiled from Schwerdtfeger (1984).

Table A2.7. *Radiation at the top of the atmosphere as a function of latitude and time of year*

Each monthly value is in MJ m$^{-2}$ d$^{-1}$ for one or mean of two dates per month given by List (1951).

| Latitude | Month | | | | | | | | | | | | Total (MJ y$^{-1}$) |
|---|---|---|---|---|---|---|---|---|---|---|---|---|---|
| | J | F | M | A | M | J | J | A | S | O | N | D | |
| 0 | 36 | 37 | 37 | 36 | 34 | 33 | 33 | 35 | 37 | 37 | 36 | 35 | 13 037 |
| 10 °S | 39 | 39 | 37 | 34 | 31 | 29 | 29 | 33 | 36 | 38 | 39 | 39 | 12 858 |
| 20 °S | 42 | 39 | 35 | 31 | 26 | 24 | 25 | 29 | 35 | 38 | 41 | 42 | 12 322 |
| 30 °S | 43 | 39 | 32 | 27 | 21 | 18 | 19 | 25 | 32 | 37 | 41 | 44 | 11 458 |
| 40 °S | 43 | 37 | 29 | 23 | 16 | 12 | 14 | 20 | 28 | 34 | 41 | 44 | 10 300 |
| 50 °S | 43 | 35 | 24 | 17 | 10 | 7 | 8 | 14 | 24 | 31 | 40 | 44 | 8 914 |
| 60 °S | 42 | 31 | 19 | 11 | 4 | 2 | 3 | 8 | 18 | 27 | 38 | 44 | 7 417 |
| 70 °S | 42 | 28 | 13 | 5 | 1 | —[a] | — | 3 | 13 | 22 | 37 | 45 | 6 174 |
| 80 °S | 44 | 26 | 6 | — | — | — | — | — | 6 | 18 | 38 | 47 | 5 592 |
| 90 °S | 44 | 26 | — | — | — | — | — | — | — | 18 | 39 | 48 | 5 411 |

[a] — = <1 MJ d$^{-1}$.

Table A2.8. *Average number of hours of daylight (sunrise to sunset) per day as a function of latitude*

Interpolated and averaged from the tables and graphs in List (1951). Refraction at low solar angles may extend these daylight estimates.

| Latitude | Month | | | | | | | | | | | |
|---|---|---|---|---|---|---|---|---|---|---|---|---|
|  | J | F | M | A | M | J | J | A | S | O | N | D |
| 0 | 12 | 12 | 12 | 12 | 12 | 12 | 12 | 12 | 12 | 12 | 12 | 12 |
| 10 °S | 13 | 12 | 12 | 12 | 12 | 12 | 12 | 12 | 12 | 12 | 12 | 12 |
| 20 °S | 13 | 13 | 12 | 12 | 11 | 11 | 11 | 12 | 12 | 13 | 13 | 13 |
| 30 °S | 14 | 13 | 12 | 11 | 11 | 10 | 10 | 11 | 12 | 13 | 14 | 14 |
| 40 °S | 15 | 14 | 12 | 11 | 10 | 9 | 10 | 11 | 12 | 13 | 14 | 15 |
| 50 °S | 16 | 14 | 12 | 11 | 9 | 8 | 8 | 10 | 12 | 14 | 16 | 16 |
| 60 °S | 18 | 16 | 13 | 10 | 8 | 6 | 7 | 9 | 12 | 15 | 17 | 19 |
| 65 °S | 20 | 16 | 13 | 10 | 6 | 4 | 5 | 8 | 12 | 15 | 18 | 22 |
| 70 °S | 24 | 18[a] | 13 | 9 | 4[a] | 0 | 2 | 7 | 12 | 16[a] | 22 | 24 |
| 75 °S | 24 | 22 | 14[a] | 8[a] | 1 | 0 | 0 | 4[a] | 12[a] | 18[a] | 24 | 24 |
| 80 °S | 24 | 24 | 15[b] | 4[a] | 0 | 0 | 0 | 1[a] | 11[a] | 21[a] | 24 | 24 |
| 85 °S | 24 | 24 | 18[b] | 2[a] | 0 | 0 | 0 | 0[a] | 10[b] | 24[a] | 24 | 24 |
| 90 °S | 24 | 24 | 23[b] | 0 | 0 | 0 | 0 | 0 | 8[b] | 24 | 24 | 24 |

[a] More than six hours difference in daylength between the first and last day of the month.
[b] More than 12 hours difference in daylength between the first and last day of the month.

Table A2.9. *Maximum solar angle (degrees) as a function of latitude*

| | Date | | | |
|---|---|---|---|---|
| Latitude | Dec 22 | Mar 22 | Jun 22 | Sep 22 |
| 0 | 67 | 90 | 67 | 90 |
| 10 °S | 77 | 80 | 57 | 80 |
| 20 °S | 88 | 70 | 47 | 70 |
| 30 °S | 86 | 60 | 37 | 60 |
| 40 °S | 74 | 50 | 27 | 50 |
| 50 °S | 64 | 40 | 17 | 40 |
| 60 °S | 54 | 30 | 7 | 30 |
| 70 °S | 44 | 20 | — | 20 |
| 80 °S | 34 | 10 | — | 10 |
| 90 °S | 24 | 0 | — | 0 |

Calculated from the algorithm given by List (1951).

Table A2.10. *Optical mass of air at low solar angles (the relative length of the path traversed by the sun's rays reaching the earth)*

Solar angle $= (90 - i)$ where $i$ is the zenith distance.

| | Solar angle (degrees) | | | | | | |
|---|---|---|---|---|---|---|---|
| | 1 | 5 | 10 | 20 | 30 | 60 | 90 |
| Optical mass | 26.96 | 10.39 | 5.60 | 2.90 | 2.00 | 1.15 | 1.00 |

Compiled from List (1951).

Table A2.11. *Albedo of various materials in Antarctica* (%)

| Material | Albedo | Location | Reference |
|---|---|---|---|
| *Ice* | | | |
| blue ice | 48 | McMurdo Ice Shelf | Paige (1968) |
| pulverised ice | 76 | McMurdo Ice Shelf | Paige (1968) |
| *Lake ice* | | | |
| smooth, blue (Nov) | 20–40 | Lake Vanda | Goldman *et al.* (1967) |
| rough, white (Dec/Jan) | 57 | Lake Vanda | Goldman *et al.* (1967) |
| dry, late season (Feb) | 34 | Lake Vanda | Goldman *et al.* (1967) |
| Feb 1135 h | 37–40 | Lake Vanda | Goldman *et al.* (1967) |
| Feb 1845 h | 56 | Lake Vanda | Goldman *et al.* (1967) |
| white, bubbly (2 years old) | 56 | Skua Lake | Goldman *et al.* (1972) |
| smooth ice | 46 | Skua Lake | Goldman *et al.* (1972) |
| dark, recently refrozen | 39 | Skua Lake | Goldman *et al.* (1972) |
| dark, recently refrozen (with few bubbles) | 14 | Skua Lake | Goldman *et al.* (1972) |
| *Water* | | | |
| at lake edge | 17 | Skua Lake | Goldman *et al.* (1972) |
| *Snow* | | | |
| fresh snow | 86 | McMurdo Ice Shelf | Paige (1968) |
| fresh snow | 80–90 | various sites | Walton (1984) |
| melting snow | 30–65 | various sites | Walton (1984) |
| on sea ice | 92 | McMurdo Sound | Palmisano *et al.* (1987b) |
| *Bare ground* | | | |
| ice-free ground | | | |
| annual mean | 25 | Bunger Oasis | Rusin (1964) |
| Jan | 16 | Bunger Oasis | Rusin (1964) |
| rock | 14 | Ross Island | Bull (1966) |
| ground | 20 | Dry Valleys | Thompson *et al.* (1971) |

# References

Abyzov, S. S., Filippova, S. N. & Kuznetsov, V. D. (1983). *Nocardiopsis antarcticus*–isolated from a glacial thickness in central Antarctica. *Bulletin of the Academy of Sciences of the USSR*, **4**, 559–68.

Ackley, S. F., Buck, K. R. & Taguchi, S. (1979). Standing crop of algae in the sea ice of the Weddell Sea region. *Deep-Sea Research*, **26A**, 269–81.

Adams, W. A. (1978). Effects of ice cover on the solar radiation regime in Canadian lakes. *Verhandlungen der Internationale Vereinigung für Theoretische und angewandte Limnologie*, **20**, 141–9.

Ainley, D. G., Fraser, W. R., Sullivan, C. W., Torres, J. J., Hopkins, T. L. & Smith, W. O. (1986). Antarctic mesopelagic micronekton: evidence from seabirds that pack ice affects community structure. *Science*, **232**, 847–9.

Ainley, D. G. & Sullivan, C. W. (1984). AMERIEZ 1983: A summary of activities on board RV Melville and USCGC Westwind. *Antarctic Journal of the United States*, **19**, 100–3.

Akiyama, M. (1979). Some ecological and taxonomic observations on the coloured snow algae found in Rumpa and Skarvsnes, Antarctica. In *Proceedings of the Symposium on Terrestrial Ecosystems in the Syowa Station Area*, eds. T. Matsuda & T. Hoshiai, pp. 27–34. Tokyo: National Institute of Polar Research.

Allanson, B. R., Hart, R. C. & Lutjeharms, J. R. E. (1981). Observations on the nutrients, chlorophyll, and primary production of the southern ocean south of Africa. *South African Journal of Antarctic Research*, **10/11**, 3–14.

Allen, S. E., Grimshaw, H. M. & Holdgate, M. W. (1967). Factors affecting the availability of plant nutrients on an antarctic island. *Journal of Ecology*, **55**, 381–96.

Allen, T. R. (1973). Common colds in Antarctica. *Journal of Hygiene (Cambridge)*, **71**, 649–56.

Allison, I., Kerry, K. & Wright, S. (1985). Observations of water mass modification in the vicinity of an iceberg. *Iceberg Research*, **1**, 3–9.

Atlas, R. M. (1986). Fate of petroleum pollutants in arctic ecosystems. In *Arctic Water Pollution Research: Applications of Science and Technology. Water Science and Technology* **18**, eds. W. A. Bridgeo & H. R. Eisenhauer, pp. 59–67. Oxford: Pergamon Press Ltd.

Atlas, R. M., Di Menna, M. E. & Cameron, R. E. (1978). Ecological investigations of yeasts in antarctic soils. *Antarctic Research Series*, **30**, 27–34.

Atlas, R. M. & Morita, R. (1987). Bacterial communities in nearshore Arctic and Antarctic marine ecosystems. In *International Congress on Microbial Ecology*, ed. F. Megusar. Ljubljana. In press.

Azam, F., Ammerman, J. W. & Cooper, N. (1981). Bacterioplankton distributional patterns and metabolic activities in the Scotia Sea. *Antarctic Journal of the United States*, **16**, 164–5.

Azam, F., Beers, J. R., Campbell, L., Carlucci, A. F., Holm-Hansen, O., Reid, F. M. H. & Karl, D. M. (1979). Occurrence and metabolic activity of organisms under the Ross Ice Shelf, Antarctica at Station J9. *Science*, **203**, 451–3.

Azov, Y. (1986). Seasonal patterns of phytoplankton productivity and abundance in nearshore oligotrophic waters of the Levant Basin (Mediterranean). *Journal of Plankton Research*, **8**, 41–53.

Badger, M. R. & Collatz, G. J. (1977). Studies on the kinetic mechanism of ribulose-1,5-bisphosphate carboxylase and oxygenase reactions with particular reference to the effect of temperature on kinetic parameters. *Carnegie Institution of Washington Yearbook*, **76**, 355–61.

Bailey, A. D. & Wynn-Williams, D. D. (1982). Soil microbiological studies at Signy Island, South Orkney Islands. *British Antarctic Survey Bulletin*, **51**, 167–91.

Barber, R. T. & Smith, R. L. (1981). Coastal upwelling ecosystems. In *Analysis of Marine Ecosystems*, ed. A. R. Longhurst, pp. 31–68. London–New York: Academic Press.

Barkov, N. I. (1985). *Ice Shelves of Antarctica*. Rotterdam: A. A. Balkema.

Barrett, P. J., Pyne, A. R. & Ward, B. L. (1983). Modern sedimentation in McMurdo Sound, Antarctica. In *Antarctic Earth Science*, ed. R. L. Oliver, P. R. James & J. B. Jago, pp. 550–4. Canberra: Australian Academy of Science.

Baust, J. G. & Lee, R. E. (1980). Environment 'homeothermy' in an antarctic insect. *Antarctic Journal of the United States*, **15**, 170–2.

Bayly, I. A. E. (1986). Ecology of a meromictic antarctic lagoon with special reference to *Drepanopus bispinosus* (Copepoda: Calanoida). *Hydrobiologia*, **140**, 199–231.

Bayly, I. A. E. & Eslake, D. (1988). Vertical distributions of a planktonic harpacticoid and a calanoid (Copepoda) in a meromictic Antarctic lake. In *High Latitude Limnology. Developments in Hydrobiology*, ed. W. F. Vincent & J. C. Ellis-Evans. The Hague: Dr Junk. In press.

Becker, E. W. (1982). Physiological studies on antarctic *Prasiola crispa* and *Nostoc commune* at low temperatures. *Polar Biology*, **1**, 99–104.

Ben-Amotz, A. (1975). Adaptation of the unicellular alga *Dunaliella parva* to a saline environment. *Journal of Phycology*, **11**, 50–4.

Benninghoff, W. S. & Benninghoff, A. S. (1985). Wind transport of electrostatically charged particles and minute organisms in Antarctica. In *Antarctic Nutrient Cycles and Food Webs*, ed. W. R. Siegfried, P. R. Condy & R. M. Laws, pp. 592–6. Berlin: Springer-Verlag.

Benninghoff, W. S. & Bonner, W. N. (1984). *Man's Impact on the Antarctic Environment: A Procedure for Evaluating Impacts from Scientific and Logistic Activities*. Cambridge: Scientific Committee on Antarctic Research, Scott Polar Institute.

Benoit, R. E. & Hall, C. L. (1970). The microbiology of some dry valley soils of Victoria Land, Antarctica. In *Antarctic Ecology*, ed. M. W. Holdgate, pp. 697–701. London: Academic Press.

Berkman, P. A., Marks, D. S. & Shreve, G. P. (1986). Winter sediment resuspension in McMurdo Sound, Antarctica, and its ecological implications. *Polar Biology*, **6**, 1–3.

Biggs, D. C. (1978). Non-biogenic fixed nitrogen in Antarctic surface waters. *Nature*, **276**, 96–7.

Biggs, D. C. (1982a). Ross Sea ammonium flux experiment. *Antarctic Journal of the United States*, **17**, 144–6.

Biggs, D. C. (1982b). Zooplankton excretion and $NH_4^+$ cycling in near-surface waters of the Southern Ocean. 1. Ross Sea, austral summer 1977–1978. *Polar Biology*, **1**, 55–67.

Biggs, D. C., Amos, A. F. & Holm-Hanson, O. (1985). Oceanographic studies of epi-pelagic ammonium distributions: the Ross Sea $NH_4^+$ flux experiment. In *Antarctic Nutrient Cycles and Food Webs*, ed. W. R. Siegfried, P. R. Condy & R. M. Laws, pp. 93–103. Berlin: Springer-Verlag.

Bird, D. F. & Kalff, J. (1986). Bacterial grazing by planktonic lake algae. *Science*, **231**, 493–5.

Block, W. (1984). Terrestrial microbiology, invertebrates and ecosystems. In *Antarctic Ecology*, ed. R. M. Laws, pp. 163–236. London: Academic Press.

Booth, B. C. & Marchant, H. J. (1987). Parmales, a new order of marine chrysophytes, with descriptions of three new genera and seven new species. *Journal of Phycology*, **23**, 245–60.

Boyd, W. L. (1967). Ecology and physiology of soil micro-organisms in polar regions. *Japanese Antarctic Research Expedition, Scientific Reports*, **Special issue No. 1**, 265–75.

Boyd, W. L. & Boyd, J. W. (1963). Soil microorganisms of the McMurdo Sound area, Antarctica. *Applied Microbiology*, **11**, 116–21.

Bradford, J. M. (1978). Sea ice organisms and their importance to the antarctic ecosystem. *New Zealand Antarctic Record*, **1**, 43–50.

Brady, H. T. (1980). 'Palaeoenvironmental and biostratigraphic studies in the McMurdo and Ross Sea regions.' Unpublished Ph.D. thesis, Macquarie University, Australia.

Brady, H. T. & Batts, B. (1981). Large salt beds on the surface of the Ross Ice Shelf near Black Island, Antarctica. *Journal of Glaciology*, **27**, 11–18.

Brady, H. T. & Martin, H. (1979). Ross Sea region in the middle miocene: a glimpse into the past. *Science*, **203**, 437–8.

Broady, P. A. (1979a). The Signy Island terrestrial reference sites: IX. The ecology of the algae of site 2, a moss carpet. *British Antarctic Survey Bulletin*, **47**, 13–30.

Broady, P. A. (1979b). Feeding studies on the collembolan *Cryptopygus antarcticus* Willem at Signy Island, South Orkney Islands. *British Antarctic Survey Bulletin*, **48**, 37–46.

Broady, P. A. (1979c). Wind dispersal of terrestrial algae at Signy Island, South Orkney Islands. *British Antarctic Survey Bulletin*, **48**, 99–102.

Broady, P. A. (1979d). A preliminary survey of the terrestrial algae of the Antarctic Peninsula and South Georgia. *British Antarctic Survey Bulletin*, **48**, 47–70.

Broady, P. A. (1979e). The terrestrial algae of Signy Island, South Orkney Islands. *British Antarctic Survey Scientific Reports*, **98**, 117p.

Broady, P. A. (1981a). Non-marine algae of Cape Bird, Ross Island and Taylor Valley, Victoria Land, Antarctica. *Melbourne University Programme in Antarctic Studies Report*, **37**, 97p.

Broady, P. A. (1981b). The ecology of chasmolithic algae at coastal locations of Antarctica. *Phycologia*, **20**, 259–72.

Broady, P. A. (1981c). Ecological and taxonomic investigations on subaerial epilithic algae from Princess Elizabeth Land and Mac. Robertson Land, Antarctica. *British Phycological Journal*, **16**, 257–66.

Broady, P. A. (1981d). The ecology of sublithic terrestrial algae at the Vestfold Hills, Antarctica. *British Phycological Journal*, **16**, 231–40.

Broady, P. A. (1982a). Taxonomy and ecology of algae in a freshwater stream in Taylor Valley, Victoria Land, Antarctica. *Archiv für Hydrobiologie Supplement*, **63**, 331–9.

Broady, P. A. (1982b). Ecology of non-marine algae at Mawson Rock, Antarctica. *Nova Hedwigia*, **36**, 209–29.

Broady, P. A. (1983). The antarctic distribution and ecology of the terrestrial chlorophytan alga *Prasiococcus calcarius* (Boye Petersen) Vischer. *Polar Biology*, **1**, 211–16.

Broady, P. A. (1984a). The vegetation of Cape Bird, Ross Island, Antarctica. *Melbourne University Programme in Antarctic Studies*, **62**.

Broady, P. A. (1984b). Taxonomic and ecological investigations of algae on steam-warmed soil on Mt Erebus, Ross Island, Antarctica. *Phycologia*, **23**, 257–71.

Broady, P. A. (1985). A preliminary report of phycological studies in Northern Victoria Land and on Ross Island during 1984–85. *Melbourne University Programme in Antarctic Studies Report*, **66**, 145 p.

Broady, P. A. (1986). Ecology and taxonomy of the terrestrial algae of the Vestfold Hills. In *Antarctic Oasis Terrestrial Environments and History of the Vestfold Hills*, ed. J. Pickard, pp. 165–202. North Ryde: Academic Press Australia.

Broady, P. A. (1987). The morphology, distribution and ecology of *Pseudococcomyxa simplex* (Mainx) Fott (Chlorophyta, Chlorellaceae), a widespread terrestrial antarctic alga. *Polar Biology*, **7**, 25–30.

Broady, P. A., Given, D., Greenfield, L. & Thompson, K. (1987). The biota and environment of fumeroles on Mt Melbourne, northern Victoria Land. *Polar Biology*, **7**, 97–113.

Bromley, A. M. (1985). Weather observations in Wright Valley, Antarctica. *New Zealand Meteorological Service Information Publication*, **11**, 37 p.

Brown, P. C. & Field, J. C. (1986). Factors limiting phytoplankton production in a nearshore upwelling area. *Journal of Plankton Research*, **8**, 55–68.

Brown, R. M., Larsen, D. A. & Bold, H. C. (1964). Airborne algae, their abundance and heterogeneity. *Science*, **143**, 583–5.

Bryden, H. L. & Pillsbury, R. D. (1977). Variability of deep flow in the Drake Passage from year-long current measurements. *Journal of Physical Oceanography*, **7**, 803–10.

Buck, K. R. & Garrison, D. L. (1983). Protists from the ice-edge region of the Weddell Sea. *Deep-Sea Research*, **30**, 1261–77.

Buinitsky, V. Kh. (1977). Organic life in sea ice. In *Polar Oceans*, ed. M. J. Dunbar, pp. 301–6. Canada Arctic Institute of North America.

Bull, C. (1966). Climatological observations in ice-free areas of southern Victoria Land, Antarctica. *Antarctic Research Series*, **9**, 177–94.

Bunt, J. S. (1963). Diatoms of Antarctic sea-ice as agents of primary production. *Nature*, **199**, 1255–7.

Bunt, J. S. (1964). Primary productivity under sea ice in Antarctic waters. 2. Influence of light and other factors on photosynthetic activities of Antarctic marine microalgae. In *Biology of the Antarctic Seas. Antarctic Research Series, 1*, ed. M. O. Lee, pp. 27–31. Washington DC: American Geophysical Union.

Bunt, J. S. (1968). Some characteristics of microalgae isolated from Antarctic sea ice. *Antarctic Research Series*, **11**, 1–14.

Bunt, J. S. & Lee, C. C. (1970). Seasonal primary production in the Antarctic sea ice at McMurdo Sound in 1967. *Journal of Marine Research*, **28**, 304–20.

Bunt, J. S. & Lee, C. C. (1972). Data on the composition and dark survival of four sea-ice microalgae. *Limnology and Oceanography*, **17**, 458–61.

Bunt, J. S. & Wood, E. J. F. (1963). Microalgae and Antarctic sea ice. *Nature*, **199**, 1254–5.

Burch, M. D. (1987). Annual cycle of phytoplankton in Ace Lake, an ice-covered saline meromictic lake. In *Proceedings of the Symposium on the Biology of the Vestfold Hills, Antarctica*, eds. G. W. Johnstone, H. R. Burton & I. A. E. Bayly. The Hague: Dr. W. Junk. In press.

Burch, M. D. & Marchant, H. J. (1983). Motility and microtubule stability of antarctic algae at sub-zero temperatures. *Protoplasma*, **115**, 240–2.

Burke, C. M. & Burton, H. R. (1988). The ecology of photosynthetic bacteria in Burton Lake, Antarctica. In *Proceedings of the Symposium on the Biology of the Vestfold Hills, Antarctica*, eds. G. W. Johnstone, H. R. Burton & I. A. E. Bayly. The Hague: Dr. W. Junk. In press.

Burkholder, P. R. & Mandelli, E. F. (1965). Productivity of microalgae in antarctic sea ice. *Science*, **149**, 872–4.

Burton, H. R. & Campbell, P. J. (1980). The climate of the Vestfold Hills, Davis Station, Antarctica, with a note on its effect on the hydrology of hypersaline Deep Lake. *Australian National Antarctic Research Expeditions Scientific Reports*, **129**, 50 p.

Cameron, R. E. (1971). Antarctic soil microbial and ecological investigations. In *Research in the Antarctic*, eds. L. O. Quam & H. D. Porter, pp. 137–89. Wash. DC: American Association for the Advancement of Science.

Cameron, R. E. (1972a). Farthest south algae and associated bacteria. *Phycologia*, **11**, 133–9.

Cameron, R. E. (1972b). Pollution and conservation of the Antarctic terrestrial ecosystem. In *Proceedings of the Colloquium on Conservation Problems in Antarctica*, ed. B. C. Parker, pp. 267–305. Lawrence, Kansas: Allen Press Inc.

Cameron, R. E. & Benoit, R. E. (1970). Microbial and ecological investigations of recent cinder cones, Deception Island, Antarctica – a preliminary report. *Ecology*, **51**, 802–9.

Cameron, R. E., Morelli, F. A. & Johnson, R. M. (1972). Bacterial species in soil and air of the Antarctic Continent. *Antarctic Journal of the United States*, **7**, 187–9.

Campbell, I. B. & Claridge, G. G. C. (1981). Soil research in the Ross Sea region of Antarctica. *Journal of the Royal Society of New Zealand*, **11**, 401–10.

Campbell, I. B. & Claridge, G. G. C. (1982). The influence of moisture on the development of soils of the cold deserts of Antarctica. *Geoderma*, **28**, 221–38.

Canfield, D. E. & Green, W. J. (1985). The cycling of nutrients in a closed-basin Antarctic lake: Lake Vanda. *Biogeochemistry*, **1**, 233–56.

Cathey, D. D., Parker, B. C., Simmons, G. M., Yongue, W. H. & Van Brunt, M. R. (1981). The microfauna of algal mats and artificial substrates in southern Victoria Land of Antarctica. *Hydrobiologia*, **85**, 3–15.

Chinn, T. J. H. (1981). Hydrology and climate in the Ross Sea area. *Journal of the Royal Society of New Zealand*, **11**, 373–86.

Chinn, T. J. & McSaveney, M. J. (1987). On the flooding of Vanda Station. *New Zealand Antarctic Record*, **7**, 23–31.

Christie, P. (1987). Nitrogen in two contrasting antarctic bryophyte communities. *Journal of Ecology*, **75**, 73–93.

Clarke, D. B. & Ackley, S. F. (1983). Relative abundance of diatoms in Weddell Sea pack ice. *Antarctic Journal of the United States*, **18**, 181–2.

Clarke, D. B. & Ackley, S. F. (1984). Sea ice structure and biological activity in the antarctic marginal ice zone. *Journal of Geophysical Research*, **89**, 2087–95.

Clarke, K. J. & Leeson, E. A. (1985). Plasmalemma structure in freezing tolerant unicellular algae. *Protoplasma*, **129**, 120–6.

Claridge, G. G. C. & Campbell, I. B. (1977). The salts in Antarctic soils, their distribution and relationship to soil processes. *Soil Science*, **123**, 337–84.

Claridge, G. G. C. & Campbell, I. B. (1985). Physical geography – soils. In *Antarctica*, ed. N. Bonner & D. Walton, pp. 62–70. USA: Pergamon Press.

Clifford, M. A. (1983). A descriptive study of the zonation of the Antarctic Circumpolar Current and its relation to wind stress and ice cover. M.S. Thesis, Texas A & M University, College Station, Texas.

Colbeck, G. (1977). Hydrographic project, Davis, 1976. *ANARE Technical Memorandum*, **66**, 35 p. Hobart: Australian Department of Science and Technology.

Collos, Y. & G. Slawyk. (1986). $^{13}C$ and $^{15}N$ uptake by marine phytoplankton-IV. Uptake ratios and the contribution of nitrate to the productivity of Antarctic waters (Indian Ocean sector). *Deep-Sea Research*, **33**, 1039–51.

Comiso, J. C. & Sullivan, C. W. (1986). Satellite microwave and in situ observations of the Weddell Sea ice cover and its marginal ice zone. *Journal of Geophysical Research*, **91**, 9663–81,

Copin-Montegut, C. & Copin-Montegut, G. (1978). The chemistry of particulate matter from the South Indian and Antarctic Oceans. *Deep-Sea Research*, **25**, 911–31.

Cota, G. F. (1985). Photoadaptation of high Arctic ice algae. *Nature*, **315**, 219–22.

Cox, G. F. N. & Weeks, W. F. (1974). Salinity variations in sea ice. *Journal of Glaciology*, **13**, 109–20.

Daniel, R., Morgan, H. & Hudson, J. A. (1987). Superbugs spring from hot water. *New Scientist*, **19 Feb.**, 36–40.

Dartnall, H. J. G. & Hollowday, E. D. (1985). Antarctic rotifers. *British Antarctic Survey Scientific Reports*, *100*, 46 p.

Davey, A. (1986). Nitrogen fixation by cyanobacteria in the Vestfold Hills. In *Antarctic Oasis*, ed. J. Pickard, pp. 203–20. North Ryde: Academic Press Australia.

Davey, A. & Marchant, H. J. (1983). Seasonal variation in nitrogen fixation by *Nostoc commune* Vaucher at the Vestfold Hills, Antarctica. *Phycologia*, **22**, 337–85.

Davis, R. C. (1980). Peat respiration and decomposition in Antarctic terrestrial moss communities. *Biological Journal of the Linnaen Society*, **14**, 39–49.

Davis, R. C. (1981). Structure and function of two antarctic terrestrial moss communities. *Ecological Monographs*, **51**, 125–43.

Davis, R. C. (1986). Environmental factors influencing decomposition rates in two antarctic moss communities. *Polar Biology*, **5**, 95–103.

Dawson, R., Schramm, W. & Bolter, M. (1985). Factors influencing the production, decomposition and distribution of organic and inorganic matter in Admiralty Bay, King George Island. In *Antarctic Nutrient Cycles and Food Webs*, eds W. R. Siegfried, P. R. Condy & R. M. Laws, pp. 109–14. Berlin: Springer-Verlag.

Dayton, P. K. & Martin, S. (1971). Observations of ice stalactites in McMurdo Sound, Antarctica. *Journal of Geophysical Research*, **76**, 1595.

Dayton, P. K. & Oliver, J. S. (1977). Antarctic soft-bottom benthos in oligotrophic and eutrophic environments. *Science*, **197**, 55–8.

Dayton, P. K., Robilliard, G. A. & De Vries, A. L. (1969). Anchor ice formation in McMurdo Sound, Antarctica and its biological effects. *Science*, **163**, 273–4.

Dayton, P. K., Robilliard, G. A., Paine, R. T. & Dayton, L. B. (1974). Biological accommodation in the benthic community at McMurdo Sound, Antarctica. *Ecological Monographs*, **44**, 105–28.

Dayton, P. K., Watson, D., Palmisano, A., Barry, J. P., Oliver, J. S. & Rivera, D. (1986). Distribution patterns of benthic microalgal standing stock at McMurdo Sound, Antarctica. *Polar Biology*, **6**, 207–13.

Deacon, G. E. R. (1982). Physical and biological zonation in the Southern Ocean. *Deep-Sea Research*, **29**, 1–15.

Deacon, G. E. R. (1984). *The Antarctic Circumpolar Ocean*. Cambridge: Cambridge University Press.

Debenham, F. (1920). A new mode of transportation by ice: The raised marine muds of south Victoria Land (Antarctica). *Quarterly Journal of the Geological Society of London*, **75**, 51–76.

De Laca, T. E. (1982a). Use of dissolved amino acids by foraminifera. *American Zoologist*, **22**, 683–90.

De Laca, T. E. (1982b). Biology and ecology of shallow-water rhizopodia in McMurdo Sound. *Antarctic Journal of the United States*, **17**, 159–60.

De Laca, T. E., Lipps, J. H. & Hessler, R. R. (1980). The morphology and ecology of a new large agglutinated antarctic foraminifer (Tertulariina: Notodendrodidae nov.) *Zoological Journal of the Linnaen Society*, **69**, 205–44.

Delille, D. & Cahet, G. (1985). Heterotrophic processes in a Kerguelen mussel-bed. In *Antarctic Nutrient Cycles and Food Webs*, eds. W. R. Siegfried, P. R. Condy & R. M. Laws, pp. 128–35. Berlin: Springer-Verlag.

Dell, R. K. (1972). Antarctic benthos. *Advances in Marine Biology*, **10**, 1–216.

De Master, D. J. (1981). The supply and accumulation of silica in the marine environment. *Geochimica et Cosmochimica Acta*, **45**, 1715–32.

Dick, E. C., Gavinski, S. S., Mahl, M. C. & Walter, G. R. (1979). A virucidal handkerchief for helping prevent transmission of respiratory infection at McMurdo Station and Scott Base during the winter fly-in period. *Antarctic Journal of the United States*, **14**, 189–90.

Dick, E. C., Jennings, L. C., Meschievitz, C. K., MacMillan, D. & Goodrum, J. (1980). Possible modification of the normal winter fly-in respiratory disease outbreak at McMurdo Station. *Antarctic Journal of the United States*, **15**, 173–4.

Dick, E. C., Polyak, F., Kapitan, K. S., Warshauer, D. M., Mandel, A. D., Thomas, B. S. & Rankin, J. (1978). Respiratory virus transmission at McMurdo Station and Scott Base (New Zealand) during the winter fly-in period 1977. *Antarctic Journal of the United States*, **13**, 170–1.

Dieckmann, G., Hemleben, C. & Spindler, M. (1987). Biogenic and mineral inclusions in a green iceberg from the Weddell sea, Antarctica. *Polar Biology*, **7**, 31–3.

Dieckmann, G., Rohardt, G., Hellmer, H. & Kipfstuhl, J. (1986). The occurrence of ice platelets at 250 m depth near the Filchner Ice Shelf and its significance for sea ice biology. *Deep-Sea Research*, **33**, 141–8.

Di Menna, M. E. (1966). Yeasts in antarctic soils. *Antonie van Leeuwenhoek*, **32**, 29–38.

Dixon, W. L., Franks, F. & Rees, T. (1981). Cold liability of phosphofructokinase from potato tubers. *Phytochemistry*, **20**, 969–72.

Dodge, C. W. (1973). *Lichen Flora of the Antarctic Continent and Adjacent Islands*. Canaan, New Hampshire: Phoenix Publishing.

Doronin, Y. P. & Kheisin, D. E. (1977). *Sea Ice*. New Delhi: Amerind Publishing Company Pvt Ltd.

Drewry, D. J., Jordan, S. R. & Jankowski, E. (1982). Measured properties of the antarctic ice sheet: surface configuration, ice thickness, volume and bedrock characteristics. *Annals Glaciology*, **3**, 83–91.

Dugdale, R. C. (1976). Nutrient cycles. In *The Ecology of the Seas* ed. D. H. Cushing & J. J. Walsh, pp. 141–72. Saunders.

Dunbar, R. B., Dehn, M. & Leventer, A. (1984). Distribution of biogenic components in surface sediments from the antarctic continental shelf. *Antarctic Journal of the United States*, **19**, 126–8.

Dunbrovin, L. I. (1979). Major types of Antarctic ice shores. *Polar Geography*, **3**, 69–75.

Edmondson, W. T. (1972). The present condition of Lake Washington. *Verhandlungen der Internationale Vereinigung für Theoretische und Angewandte Limnologie,* **18**, 284–91.

Ellis-Evans, J. C. (1981a). Fresh water biology in the Antarctic: II Microbial numbers and activity in nutrient-enriched Heywood Lake, Signy Island. *British Antarctic Survey Bulletin*, **54**, 105–21.

Ellis-Evans, J. C. (1981b). Freshwater microbiology in the Antarctic: I. Microbial numbers and activity in oligotrophic Moss Lake, Signy Island. *British Antarctic Survey Bulletin*, **54**, 85–104.

Ellis-Evans, J. C. (1982). Seasonal microbial activity in Antarctic freshwater lake sediments. *Polar Biology*, **1**, 129–40.

Ellis-Evans, J. C. (1984). Methane in maritime Antarctic freshwater lakes. *Polar Biology*, **3**, 63–71.

Ellis-Evans, J. C. (1985a). Fungi from maritime antarctic freshwater environments. British Antarctic Survey Bulletin, **68**, 37–45.

Ellis-Evans, J. C. (1985b). Decomposition processes in maritime antarctic lakes. In *Antarctic Nutrient Cycles and Food Webs*, ed. W. R. Siegfried, P. R. Condy & R. M. Laws, pp. 253–60. Berlin: Springer-Verlag.

Ellis-Evans, J. C. (1985c). Interactions of bacterio- and phyto-plankton in nutrient cycling within eutrophic Heywood Lake, Signy Island. In *Antarctic Nutrient Cycles and Food Webs*, ed. W. R. Siegfried, P. R. Condy & R. M. Laws, pp. 261–4.

Ellis-Evans, J. C. & Lemon, E. C. G. (1988). The significance of iron to nutrient cycling in certain maritime freshwater lakes. In *High Latitude Limnology. Developments in Hydrobiology*, ed. W. F. Vincent & J. C. Ellis-Evans, in press. The Hague: Dr. W. Junk.

Ellis-Evans, J. C. & Wynn-Williams, D. D. (1985). The interaction of soil and lake microflora at Signy Island. In *Antarctic Nutrient Cycles and Food webs*, ed. W. R. Siegfried, P. R. Condy & R. M. Laws, pp. 662–8. Berlin: Springer-Verlag.

El-Sayed, S. Z. (1967). On the productivity of the southwest Atlantic Ocean and the waters west of the Antarctic Peninsula. In *Biology of the Antarctic Seas III. Antarctic Research Series, 11*, ed. W. Schmitt & G. Llano, pp. 15–47. Washington DC: American Geophysical Union.

El-Sayed, S. Z. (1971). Biological aspects of the pack ice ecosystem. In *Symposium on Antarctic Ice and Water Masses*, ed. G. Deacon, pp. 35–54. Tokyo: Scientific Committee on Antarctic Research.

El-Sayed, S. Z. (1984). Productivity of Antarctic waters – a reappraisal. In *Marine Phytoplankton and Productivity. Lecture Notes on Coastal and Estuarine Studies, 8*, ed. O. Holm-Hansen, L. Bolis & R. Gilles, pp. 19–34. Berlin: Springer-Verlag.

El-Sayed, S. Z. & Mandelli, E. F. (1965). Primary production and standing crop of phytoplankton in the Weddell Sea and Drake Passage. In *Biology of the Antarctic Seas II. Antarctic Research Series, 5*, ed. G. A. Llano, pp. 87–106. Washington DC: American Geophysical Union.

El-Sayed, S. Z. & Taguchi, S. (1981). Primary production and standing crop of phytoplankton along the ice-edge in the Weddell Sea. *Deep-Sea Research*, **28A**, 1017–32.

El-Sayed, S. Z. & Turner, J. T. (1977). Productivity of the Antarctic and subtropical regions: A comparative study. In *Polar Oceans*, ed. M. J. Dunbar, pp. 463–503. Montreal: Arctic Institute of North America.

El-Sayed, S. Z. & Weber, L. H. (1982). Spatial and temporal variations in phytoplankton biomass and primary productivity in the South-west Atlantic and Scotia Sea. *Polar Biology*, **1**, 83–90.

Emery, W. J. (1977). Antarctic Polar Frontal Zone from Australia to the Drake Passage. *Journal of Physical Oceanography*, **7**, 811–22.

Eppley, R. W. (1972). Temperature and phytoplankton growth in the sea. *Fishery Bulletin*, **70**, 1063–85.

Erickson, A. W. (1984). Aerial census of seals, whales, and penguins in the pack ice of the northwestern Weddell sea, November 1983. *Antarctic Journal of the United States*, **19**, 121–4.

Farman, J. C., Gardiner, B. G. & Shanklin, J. D. (1985). Large losses of total ozone in Antarctica reveal seasonal $ClO_x/NO_x$ interaction. *Nature*, **315**, 207–8.

Fenchel, T. & Lee, C. C. (1972). Studies on ciliates associated with sea ice from Antarctica. 1. The nature of the fauna. *Archiv für Protistenkunde*, **114**, 5231–6.

Ferris, J. M. & Burton, H. R. (1987). The annual cycle of heat content and mechanical stability of hypersaline Deep Lake, Vestfold Hills. In *Proceedings of the Symposium on the Biology of the Vestfold Hills. Developments in Hydrobiology*, eds. G. W. Johnstone, H. R. Burton & I. A. E. Bayly, in press. The Hague: Dr W. Junk.

Fifield, R. (1985). How science may wreck antarctic ecology. *New Scientist*, **2 May**, p. 9.

Fitzsimmons, J. M. (1971). On the food habits of certain antarctic arthropods from coastal Victoria Land and adjacent islands. *Pacific Insects Monograph*, **25**, 121–5.

Fletcher, L. D., Kerry, E. J. & Weste, G. M. (1985). Microfungi of Mac. Robertson and Enderby Lands, Antarctica. *Polar Biology*, **4**, 81–8.

Fogg, G. E. (1967). Observations on the snow algae of the South Orkney Islands. *Philosophical Transactions of the Royal Society Series B*, **252**, 279–87.

Fogg, G. E. (1969). Survival of algae under adverse conditions. *Symposium of the Society for Experimental Biology*, **23**, 123–42.

Fogg, G. E., Stewart, W. D. P., Fay, P. & Walsby, A. E. (1973). *The Blue-Green Algae*. London: Academic Press.

Foster, T. D. (1984). The marine environment. In *Antarctic Ecology*, ed. R. M. Laws, pp. 345–71. London: Academic Press.

Foster, T. D. & Carmack, E. C. (1976). Temperature and salinity structure in the Weddell Sea. *Journal of Physical Oceanography*, **6**, 36–44.

Franks, F. (1985). *Biophysics and Biochemistry at Low Temperature*. Cambridge: Cambridge University Press.

Franzmann, P. D., Burton, H. R. & McMeekin, T. A. (1987a). *Halomonas subglaciescola*, a new species of halotolerant bacteria isolated from Antarctica. *International Journal of Systematic Bacteriology*, **37**, 27–34.

Franzmann, P. D., Deprez, P. P., Burton, H. R. & van den Hoff, J. (1987b). Limnology of Organic Lake, Antarctica, a meromictic lake that contains high concentrations of dimethyl sulfide. *Australian Journal of Marine and Freshwater Research*, **38**, 409–17.

Fraser, W. R. & Ainley, D. G. (1986). Ice edges and seabird occurrence in Antarctica. *BioScience*, **36**, 258–63.

Friedmann, E. I. (1978). Melting snow in the dry valleys is a source of water for endolithic microorganisms. *Antarctic Journal of the United States*, **13**, 162–3.

Friedmann, E. I. (1982). Endolithic microorganisms in the Antarctic cold desert. *Science*, **215**, 1045–53.

Friedmann, E. I. & Kibler, A. P. (1980). Nitrogen economy of endolithic microbial communities in hot and cold deserts. *Microbial Ecology*, **6**, 95–108.

Friedmann, E. I., La Rock, P. A. & Brunson, J. O. (1980). Adenosine triphosphate (ATP), chlorophyll, and organic nitrogen in endolithic microbial communities and adjacent soils in the dry valleys of southern Victoria Land. *Antarctic Journal of the United States*, **15**, 164–6.

Friedmann, E. I., McKay, C. P. & Nienow, J. A. (1987). The cryptoendolithic microbial habitat in the Ross Desert of Antarctica: satellite transmitted continuous nanoclimate data, 1984 to 1986. *Polar Biology*. In press.

Friedmann, E. I. & Ocampo-Friedmann, R. (1976). Endolithic blue-green algae in the dry valleys: primary producers in the Antarctic desert ecosystem. *Science*, **193**, 1247–9.

Friedmann, E. I. & Ocampo-Friedmann, R. (1984). Endolithic micro-organisms in extreme dry environments: Analysis of a lithobiontic microbial habitat. In *Current Perspectives in Microbial Ecology*, ed. M. J. Klug & C. A. Reddy, pp. 177–85. Washington DC: American Society for Microbiology.

Friedmann, E. I. & Weed, R. (1987). Microbial trace fossil formation, biogenous and abiotic weathering in the Antarctic cold desert. *Science*. In press.

Fritsch, F. E. (1912). Freshwater algae collected in the South Orkney Islands by R. N. R. Brown, 1902–04. *Journal of the Linnean Society of Botany*, **40**, 293–338.

Fryxell, G. A., Theriot, E. C. & Buck, K. R. (1984). Phytoplankton, ice algae, and choanoflagellates from AMERIEZ, the southern Atlantic and Indian Oceans. *Antarctic Journal of the United States*, **19**, 107-9.

Fuhrman, J. A. & Azam, F. (1980). Bacterioplankton secondary production estimates for coastal waters of British Columbia, Antarctic and California. *Applied and Environmental Microbiology*, **39**, 1085–95.

Fuhrman, J. A. & Azam, F. (1982). Thymidine incorporation as a measure of heterotrophic bacterioplankton production in marine surface waters: evaluation and field results. *Marine Biology*, **66**, 109–20.

Gallagher, J. B. (1985). The influence of iron and manganese on nutrient cycling in shallow freshwater antarctic lakes. In *Antarctic Nutrient Cycles and Food Webs*, ed. W. R. Siegfried, P. R. Condy & R. M. Laws, pp. 234–7. Berlin: Springer-Verlag.

Garrison, D. L., Ackley, S. F. & Buck, K. R. (1983a). A physical mechanism for establishing algal populations in frazil ice. *Nature*, **306**, 363–5.

Garrison, D. L., Buck, K. L. & Silver, M. W. (1983b). Studies of ice–algal communities in the Weddell Sea. *Antarctic Journal of the United States*, **18**, 179–81.

Garrison, D. L., Buck, K. R. & Silver, M. W. (1984). Microheterotrophs in the ice-edge zone. *Antarctic Journal of the United States*, **19**, 109–111.

Garrison, D. L. & Van Scoy, K. (1985). Wilkes Land Expedition 1985: Biological observations in the ice-edge zone. *Antarctic Journal of the United States*, **20**, 123–4.

Glibert, P. M., Biggs, D. C. & McCarthy, J. J. (1982). Utilisation of ammonium and nitrate during austral summer in the Scotia Sea. *Deep-Sea Research*, **29**, 837–50.

Glibert, P. M. & McCarthy, J. J. (1984). Uptake and assimilation of ammonium and nitrate by phytoplankton: indices of nutritional status for natural assemblages. *Journal of Plankton Research*, **6**, 677–97.

Goldman, C. R., Mason, D. T. & Hobbie, J. E. (1967). Two antarctic desert lakes. *Limnology and Oceanography*, **12**, 295–310.

Goldman, C. R., Mason, D. T. & Wood, B. J. B. (1963). Light injury and inhibition in antarctic freshwater phytoplankton. *Limnology and Oceanography*, **8**, 313–22.

Goldman, C. R., Mason, D. T. & Wood, B. J. B. (1972). Comparative limnology of two small lakes on Ross Island, Antarctica. In *Antarctic Terrestrial Ecology. Antarctic Research series*, *20*, ed. G. A. Llano, pp. 1–50. Washington DC: American Geophysical Union.

Gordon, A. L. (1981). Seasonality of Southern Ocean sea ice. *Journal of Geophysical Research*, **86**, 4193–7.

Gow, A. J., Ackley, S. F., Weeks, W. F. & Govoni, J. W. (1981). Physical and structural characteristics of Antarctic sea ice. Paper presented at the Third International Symposium on Antarctic Glaciology, Columbus, Ohio.

Gow, A. J. & Epstein, S. (1972). On the use of stable isotopes to trace the origins of ice in a floating ice tongue. *Journal of Geophysical Research*, **77**, 6552–7.

Grainger, E. H. (1977). The annual nutrient cycle in sea ice. In *Polar Oceans*, ed. M. J. Dunbar, pp 285–99. Canada: Arctic Institute of North America.

Green, T. G. A. (1981). *Annual Report of the University of Waikato, Antarctic Research Unit, 10.* Hamilton, New Zealand: University of Waikato.

Green, W. J., Canfield, D. E., Lee, G. F. & Jones, R. A. (1986). Mn, Fe, Cu and Cd distributions and residence times in closed basin Lake Vanda (Wright Valley, Antarctica). *Hydrobiologia*, **134**, 237–48.

Greenfield, L. (1981). Pathogenic microbes in Antarctica. *New Zealand Antarctic Record*, **3**, 38.

Greenfield, L. (1983). Thermophilic fungi and actinomycetes from Mt Erebus and a fungus pathogenic to *Bryum antarcticum* at Cape Bird. *New Zealand Antarctic Record*, **4**, 10–11.

Greenfield, L. & Wilson, G. (1981). *University of Canterbury, Antarctica Expedition No. 19 Report.* Christchurch, New Zealand: University of Canterbury. 47 p.

Gregory, M. R., Kirk, R. M. & Mabin, M. C. G. (1984). Pelagic tar, oil, plastics and other litter in surface waters of the New Zealand sector of the southern ocean, and on Ross Dependency shores. *New Zealand Antarctic Record*, **6**, 12–28.

Gribbon, P. W. F. (1979). Cryoconite holes on Sermikavasak, West Greenland. *Journal of Glaciology*, **22**, 177–81.

Grossi, S. M., Kottmeier, S. T., Moe, R. L., Taylor, G. T. & Sullivan, C. W. (1987). Sea ice microbial communities VI. Growth and production in bottom ice under graded snow cover. *Marine Ecology – Progress Series*, **35**, 153–64.

Grossi, S., Kottmeier, S. T. & Sullivan, C. W. (1984). Sea ice microbial communities. III. Seasonal abundance of microalgae and associated bacteria, McMurdo Sound, Antarctica. *Microbial Ecology*, **10**, 231–42.

Gruzov, E. N., Propp, M. V. & Pushkin, H. F. (1967). Hydrobiological diving work in the Antarctic. *Soviet Antarctic Expedition Information Bulletin*, **6**, 405–8.

Hale, M. E. (1987). Epilithic lichens in the Beacon sandstone formation, Victoria Land, Antarctica. *Lichenologist*, **19**. In press.

Hand, R. M. & Burton, H. R. (1981). Microbial ecology of an Antarctic saline meromictic lake. *Hydrobiologia*, **82**, 363–74.

Hanson, R. B., Lowery, H. K. (1985). Spatial distribution, structure, biomass, and physiology of microbial assemblages across the Southern Ocean Frontal Zones during the late austral winter. *Applied and Environmental Microbiology*, **49**, 1029–39.

Hanson, R. B., Lowery, H. K., Shafer, D., Sorocco, R. & Pope, D. H. (1983a). Microbes in antarctic waters of the Drake Passage: Vertical patterns of substrate uptake, productivity and biomass in January 1980. *Polar Biology*, **2**, 179–88.

Hanson, R. B., Shafer, D., Ryan, T., Pope, D. H. & Lowery, H. K. (1983b). Bacterioplankton in Antarctic Ocean waters during the late austral winter: abundance, frequency of dividing cells, and estimates of production. *Applied and Environmental Microbiology*, **45**, 1622–32.

Hart, T. J. (1934). On the phytoplankton of the south-west Atlantic and the Bellingshausen Sea. *Discovery Reports*, **8**, 1–268.

Hart, T. J. (1942). Phytoplankton periodicity in Antarctic waters. *Discovery Reports*, **21**: 261–356.

Hasle, G. R. (1969). An analysis of the phytoplankton of the Pacific Southern Ocean: abundance, composition and distribution during the Brategg Expedition 1947/48. *Hval Radets Skriffer*, **52**, 1–168.

Hawes, I. (1983a). Turbulence and its consequences for phytoplankton development in two ice-covered Antarctic lakes. *British Antarctic Survey Bulletin*, **60**, 69–81.

Hawes, I. (1983b). Nutrients and their effects on phytoplankton populations in lakes on Signy Island, Antarctica. *Polar Biology*, **2**, 115–26.

Hawes, I. (1985). Light climate and phytoplankton photosynthesis in maritime Antarctic lakes. *Hydrobiologia*, **123**, 69–79.

Hawes, I. (1988). Filamentous green algae in freshwater streams on Signy Island. In *High Latitude Limnology. Developments in Hydrobiology*, eds. W. F. Vincent & J. C. Ellis-Evans. The Hague: Junk Publishers. In press.

Hawthorn, G. R. & Ellis-Evans, J. C. (1984). Benthic protozoa from maritime antarctic freshwater lakes and pools. *British Antarctic Survey Bulletin*, **62**, 67–81.

Hayes, P. K., Whitaker, T. M. & Fogg, G. E. (1984). The distribution and nutrient status of phytoplankton in the Southern Ocean between 20° and 70°W. *Polar Biology*, **3**, 153–65.

Heath, R. A. (1981). Oceanic fronts around southern New Zealand. *Deep-Sea Research*, **28A**, 547–60.

Heinbokel, J. F. & Coats, D. W. (1984). Reproductive dynamics of ciliates in the ice-edge zone. *Antarctic Journal of the United States*, **19**, 111–13.

Herbert, R. A. & Bell, C. R. (1977). Growth characteristics of an obligately psychrophilic *Vibrio* sp. *Archives for Microbiology*, **113**, 215–20.

Hewes, C. D., Holm-Hansen, O. & Sakshaug, E. (1985). Alternate carbon pathways at lower trophic levels in the Antarctic food web. In *Antarctic Nutrient Cycles and Food Webs*, ed. W. R. Siegfried, P. R. Condy & R. M. Laws, pp. 277–83. Berlin: Springer-Verlag.

Heywood, R. B. (1984). Antarctic inland waters. In *Antarctic Ecology*, ed. R. M. Laws, pp. 279–344. London: Academic Press.

Heywood, R. B., Everson, I. & Priddle, J. (1985). The absence of krill from the South Georgia zone, winter 1983. *Deep-Sea Research*, **32**, 369–78.

Heywood, R. B. & Whittaker, T. M. (1984). The Antarctic marine flora. In *Antarctic Ecology*, ed. R. M. Laws, pp. 373–419. London: Academic Press.

Hirano, M. (1965). Freshwater algae in the Antarctic regions. In *Biogeography and Ecology in Antarctica*, ed. J. van Miegham & P. van Oye, pp. 127–93. The Hague: Dr W. Junk.

Hochachka, P. W. & Somero, G. N. (1984). *Biochemical Adaptation*. Princeton, N.J.: Princeton University Press.

Hodson, R. E., Azam, F., Carlucci, A. F., Fuhrman, J. A., Karl, D. M., Holm-Hansen, O. (1981). Microbial uptake of dissolved organic matter in McMurdo Sound, Antarctica. *Marine Biology*, **61**, 89–94.

Hoham, R. W. (1975). Optimum temperatures and temperature ranges for growth of snow algae. *Arctic and Alpine Research*, **7**, 13–24.

Hoham, R. W. (1980). Unicellular chlorophytes – snow algae. In *Phytoflagellates*, ed. E. R. Cox, pp. 61–84. New York: Elsevier North Holland, Inc.

Holm-Hansen, O. (1963). Viability of blue-green algae after freezing. *Physiologia plantarum*, **16**, 530–40.

Holm-Hansen, O., El-Sayed, S. Z., Franceschini, G. A. & Cuhel, R. L. (1977). Primary production and the factors controlling phytoplankton growth in the Southern Ocean. In *Adaptations within Antarctic Ecosystems*, ed. G. A. Llano, pp. 11–50. Washington: Smithsonian Institution.

Hopkins, T. L. (1985). Food web of an Antarctic midwater ecosystem. *Marine Biology*, **89**, 197–212.

Horne, A. J. (1972). The ecology of nitrogen fixation on Signy Island, South Orkney Islands. *British Antarctic Survey Bulletin*, **27**, 1–18.

Horne, A. J., Fogg, G. E. & Eagle, D. J. (1969). Studies *in situ* of the primary production of an area of inshore Antarctic sea. *Journal of the Marine Biological Association of the United Kingdom*, **49**, 393–405.

Horner, R. A. (1976). Sea ice organisms. *Oceanography and Marine Biology Annual Review*, **14**, 167–82.

Horrigan, S. G. (1981). Primary production under the Ross Ice Shelf, Antarctica. *Limnology and Oceanography*, **26**, 378–82.

Hoshiai, T. (1977). Seasonal change of ice communities in the sea ice near Syowa Station, Antarctica. In *Polar Oceans*, ed. M. J. Dunbar, pp. 307–17. Canada: Arctic Institute of North America.

Hoshiai, T. & Matsuda, T. (1979). Adélie penguin rookeries in the Lutzow-Holm Bay area and relation of rookery to algal biomass in soil. *Tokyo: Memoirs of the National Institute of Polar Research*, Special Issue, **11**, 140–52.

Howard-Williams, C. & Vincent, W. F. (1988). Microbial communities in southern Victoria Land streams (Antarctica) I. Photosynthesis. In *High Latitude Limnology. Developments in Hydrobiology*, ed. W. F. Vincent & J. C. Ellis-Evans. The Hague: Dr. W. Junk. In press.

Howard-Williams, C., Vincent, C. L., Broady, P. A. & Vincent, W. F. (1986a). Antarctic stream ecosystems: variability in environmental properties and algal community structure. *Internationale Revue der gesamten Hydrobiologie*, **71**, 511–44.

Howard-Williams, C., Vincent, W. F. & Wratt, G. S. (1986b). The Alph River ecosystem: A major freshwater environment in southern Victoria Land. *New Zealand Antarctic Record*, **7**, 21–33.

Jacka, T. H. (1983). A computer data base for antarctic sea ice extent. *Australian National Antarctic Research Expeditions Research Notes*, **13**.

Jacobs, S. S., Fairbanks, R. G. & Horibe, Y. (1985). Origin and evolution of water masses near the antarctic continental margin: evidence from $H_2\,^{18}O/H_2\,^{16}O$ ratios in seawater. In *Oceanology of the Antarctic Shelf. Antarctic Research Series, 43*, ed. S. S. Jacobs, pp. 59–85. Washington DC: American Geophysical Union.

Jacques, G. (1983). Some ecophysiological aspects of the antarctic phytoplankton. *Polar Biology*, **2**, 27–33.

Jacques, G. & Minas, M. (1981). Production primaire dans le secteur indien de l'Ocean Antarctique en fin d'été. *Oceanologica Acta*, **4**, 33–41.

Jennings, J. C., Gordon, L. I. & Nelson, D. M. (1984). Nutrient depletion indicates high primary productivity in the Weddell Sea. *Nature*, **309**, 51–4.

Joyce, T. M., Patterson, S. L. & Millard, R. C. (1981). Anatomy of a cyclonic ring in the Drake Passage. *Deep-Sea Research*, **28A**, 1265–87.

Kalff, J. & Welch, H. E. (1974). Phytoplankton production in Char Lake, a natural polar lake, and in Meretta Lake, a polluted polar lake, Cornwallis Island, Northwest Territories. *Journal of the Fisheries Research Board of Canada*, **31**, 621–36.

Kappen, L. & Friedmann, E. I. (1983). Ecophysiology of lichens in the Dry Valleys of southern Victoria Land, Antarctica II $CO_2$ gas exchange in cryptoendolithic lichens. *Polar Biology*, **1**, 227–32.

Kappen, L., Friedmann, E. I. & Garty, J. (1981). Ecophysiology of lichens in the Dry Valleys of southern Victoria Land, Antarctica. I. Microclimate of the cryptoendolithic lichen habitat. *Flora*, **171**, 216–35.

Kellogg, D. E. & Kellogg, T. B. (1984). Diatoms from the McMurdo Ice Shelf, Antarctica. *Antarctic Journal of the United States*, **19**, 76–7.

Kellogg, D. E., Kellogg, T. B., Dearborn, J. H., Edwards, K. C. & Fratt, D. B. (1982). Diatoms from brittle star stomachs: implications for sediment reworking. *Antarctic Journal of the United States*, **17**, 167–9.

Kerry, K. R., Grace, D. R., Williams, R. & Burton, H. R. (1977). Studies on some saline lakes in the Vestfold Hills, Antarctica. In *Adaptations within Antarctic Ecosystems*, ed. G. A. Llano, pp. 839–58. Washington DC: Smithsonian Institution.

Keys, J. R. (1984). *Antarctic Marine Environments and Offshore Oil*. Wellington: Commission for the Environment, New Zealand Government.

Kobori, H., Sullivan, C. W. & Shizuya, H. S. (1984a). Heat labile alkaline phosphatase from antarctic bacteria: rapid end labelling of nucleic acids. *Proceedings of the National Academy of Science*, **81**, 6691.

Kobori, H., Sullivan, C. W. & Shizuya, H. S. (1984b). Bacterial plasmids in Antarctic natural microbial assemblages. *Applied and Environmental Microbiology*, **48**, 515–18.

Kol, E. (1968). *Kryobiologie* I. *Kryovegetation*. Berlin: E. Schweizerbart'sche Verlagsbuchhandlung.

Kol, E. & Flint, E. A. (1968). Algae in green ice from the Balleny Islands, Antarctica. *New Zealand Journal of Botany*, **6**, 249–61.

Konlechner, J. C. (1985). An investigation of the fate and effects of a paraffin based crude oil in an antarctic terrestrial ecosystem. *New Zealand Antarctic Record*, **6**, 40–6.

Kottmeier, S. T., Grossi, S. M. & Sullivan, C. W. (1987). Sea ice microbial communities VIII. Bacterial production in annual sea ice of McMurdo Sound, Antarctica. *Marine Ecology – Progress Series*, **35**, 175–86.

Kovacs, A. & Gow, A. J. (1977). Subsurface measurements of the Ross Ice Shelf, McMurdo Sound, Antarctica. *Antarctic Journal of the United States*, **12**, 146–8.

Kushner, D. J. (1978). Life in high salt and solute concentrations: halophilic bacteria. In *Microbial Life in Extreme Environments*, ed. D. J. Kushner, pp. 317–68. London: Academic Press.

Le Jehan, S. (1982). Contribution à l'étude des matières organiques et nutritives dans deux écosystèmes eutrophes: Cycles de l'azote, du phosphore et du silicum. *Thèse de spécialité. UBO Brest*. 236 p.

Lee, C. C. & Fenchel, T. (1972). Studies on ciliates associated with sea-ice from Antarctica. II. Temperature responses and tolerances in ciliates from Antarctic, temperate and tropical latitudes. *Archiv fur Protistenkunde*, **114**, 237–44.

Lewis, E. L. & Weeks, W. F. (1971). Sea ice: some polar contrasts. In *Symposium on Antarctic Ice and Water Masses*, ed. G. Deacon, pp. 23–33. Tokyo: Scientific Committee on Antarctic Research.

Lewis Smith, R. I. (1984). Terrestrial plant biology of the sub-Antarctic and Antarctic. In *Antarctic Ecology*, ed. R. M. Laws, pp. 61–162. London: Academic Press.

Li, W. K. W. & Dickie, P. M. (1984). Rapid enhancement of heterotrophic but not photosynthetic activities in arctic microbial plankton at mesobiotic temperatures. *Polar Biology*, **3**, 217–26.

Lindeboom, H. J. (1984). The nitrogen pathway in a penguin rookery. *Ecology*, **65**, 269–77.

Lipps, J. H. & Krebs, W. N. (1974). Planktonic foraminifera associated with Antarctic sea ice. *Journal of Foraminiferan Research*, **4**, 80–5.

List, R. J. (1951). *Smithsonian Meteorological Tables*. Sixth edition. Washington DC: Smithsonian Institution.

Littlepage, J. L. (1965). Oceanographic investigations in McMurdo Sound, Antarctica. In *Biology of the Antarctic Seas II*, ed. G. A. Llano, pp. 1–37. Washington, DC: American Geophysical Union. *Antarctic Research Series*, 5.

Llano, G. (1972). Antarctic conservation: prospects and retrospects. In *Proceedings of the Colloquium on Conservation Problems in Antarctica*, ed. B. C. Parker, pp. 1–12. Lawrence, Kansas: Allen Press.

Lyon, G. L. & Giggenbach, W. F. (1974). Geothermal activity in Victoria Land, Antarctica. *New Zealand Journal of Geology and Geophysics*, **17**, 511–21.

McConville, M. J., Mitchell, C. & Wetherbee, R. (1985). Patterns of carbon assimilation in a microalgal community from annual sea ice, East Antarctica. *Polar Biology*, **4**, 135–41.

McConville, M. J. & Wetherbee, R. (1983). The bottom-ice microalgal community from annual ice in the inshore waters of East Antarctica. *Journal of Phycology*, **19**, 431–9.

McFadden, G. I., Moestrup, O. & Wetherbee, R. (1982). *Pyramimonas gelidicola* sp. nov. (Prasinophyceae), a new species isolated from Antarctic sea ice. *Phycologia*, **21**, 103–11.

McKay, C. P., Clow, G. D., Wharton, R. A. & Squyres, S. W. (1985). Thickness of ice on perennially frozen lakes. *Nature*, **313**, 561–2.

McKay, C. P. & Friedmann, E. I. (1985). The cryptoendolithic microbial environment in the antarctic cold desert: temperature variations in nature. *Polar Biology*, **4**, 19–25.

McLellan, M. R., Morris, G. J., Coulson, G. E., James, E. R. & Kalinina, L. V. (1984). Role of cytoplasmic proteins in cold shock injury of *Amoeba*. *Cryobiology*, **21**, 44–59.

McMurchie, E. J. & Raison, J. K. (1979). Membrane lipid fluidity and its effect on the activation energy of membrane associated enzymes. *Biochimica Biophysica Acta*, **554**, 3654–74.

Marchant, H. J. (1985). Choanoflagellates in the antarctic marine food chain. In *Antarctic Nutrient Cycles and Food Webs*, ed. W. R. Siegfried, P. R. Condy & R. M. Laws, pp. 271–6. Berlin: Springer-Verlag.

Markov, K. K., Bardin, V. I., Lebedev, V. L., Orlov, A. I. & Svetova, I. A. (1970). *The Geography of Antarctica*. Jerusalem: Israel Program for Scientific Translations.

Marra, J. & Boardman, D. C. (1984). Late winter chlorophyll *a* distributions in the Weddell Sea. *Marine Ecology – Progress Series*, **19**, 197–205.

Matthews, P. C. & Heaney, S. I. (1987). Solar heating and its influence on mixing in ice-covered lakes. *Freshwater Biology*, **18**, 135–49.

Maurette, M., Hammer, C., Brownlee, D. E., Reeh, N. & Thomsen, H. H. (1986). Placers of cosmic dust in the blue ice lakes of Greenland. *Science*, **233**, 869–72.

Maykut, G. A. (1985). The ice environment. In *Sea Ice Biota*, ed. R. A. Horner, pp. 21–82. Boca Raton, Fla.: CRC Press.

Mazur, P. (1969). Freezing injury in plants. *Annual Review of Plant Physiology*, **20**, 419–48.

Mazur, P. (1970). Cryobiology: the freezing of biological systems. *Science*, **168**, 939–49.

Megahan, W. F., Meiman, J. R. & Goodell, B. C. (1970). The effect of albedo-reducing materials on net radiation at a snow surface. *Bulletin of the International Association of Scientific Hydrology*, **15**, 69–80.

Mellor, M. & McKinnon, G. (1960). The Amery Ice Shelf and its hinterland. *Polar Record*, **10**, 30–4.

Meschievitz, C. K., Dick, E. C., Jennings, L. C., Yule, R. Prebble, M., Robinson, T., Varcoe, G., Roper, C. & Lewis, G. (1982). Possible modification of summer season respiratory disease transmission in an open community at Scott Base. *Antarctic Journal of the United States*, **17**, 198–9.

Meyer-Rochow, V. B. (1979). Kleinstlebewesen in extremen Biotopenzum 'Tuempeln' in die Antarktis. *Mikrokosmos*, **67**, 34–8.

Mikell, A. T., Parker, B. C. & Simmons, G. M. (1984). Response of an antarctic lake heterotrophic community to high dissolved oxygen. *Applied and Environmental Microbiology*, **47**, 1062–6.

Milam, R. W. & Anderson, J. B. (1981). Distribution and ecology of recent benthonic foraminifera on the Adélie–George V continental shelf and slope, Antarctica. *Marine Micropaleontology*, **6**, 297–325.

Miller, K. L. & Leschine, S. B. (1984). Halotolerant *Planococcus* from antarctic dry valley soil. *Current Microbiology*, **11**, 205–9.

Miller, K. L., Leschine, S. B. & Huguenin, R. L. (1983). Halotolerance of micro-organisms isolated from saline antarctic dry valley soils. *Antarctic Journal of the United States*, **18**, 222–3.

Miller, M. A., Krempin, D. W., Manahan, D. T. & Sullivan, C. W. (1984). Growth rates, distribution, and abundance of bacteria in the ice-edge zone of the Weddell and Scotia Seas, Antarctica. *Antarctic Journal of the United States*, **19**, 103–5.

Moe, R. L. & Silva, P. C. (1977). Antarctic marine flora: uniquely devoid of kelps. *Science*, **196**, 1206–8.

Morita, R. Y. (1975). Psychrophilic bacteria. *Bacteriological Reviews*, **39**, 144–67.

Morita, R. Y., Griffiths, R. P. & Hayasaka, S. S. (1977). Heterotrophic activity of microorganisms in Antarctic waters. In *Adaptations within Antarctic ecosystems*, ed. G. A. Llano, pp. 99–113. Washington DC: Smithsonian Institution.

Morley, J. J. & Stepien, J. C. (1984). Siliceous microfauna in waters beneath Antarctic sea ice. *Marine Ecology – Progress Series*, **19**, 207–10.

Morris, G. J. & Clarke, A. (1981). *Effects of Low Temperatures on Biological Membranes*. New York: Academic Press. 456 pp.

Mullineaux, L. S. & De Laca, T. E. (1984). Distribution of Antarctic benthic foraminifers settling on the pecten *Adamussium colbecki*. *Polar Biology*, **3**, 185–9.

Nelson, D. M. & Gordon, L. I. (1982). Production and pelagic dissolution of biogenic silica in the Southern Ocean. *Geochimica et Cosmochimica Acta*, **46**, 491–501.

Nelson, D. M., Gordon, L. I. & Smith, W. O. (1984). Phytoplankton dynamics of the marginal ice zone of the Weddell Sea, November and December, 1983. *Antarctic Journal of the United States*, **19**, 105–7.

Nelson, D. M. & Smith, W. O. (1986). Phytoplankton bloom dynamics of the western Ross Sea ice edge-II. Mesoscale cycling of nitrogen and silica. *Deep-Sea Research*, **33**, 1389–1412.

Nelson, K. H. & Thompson, T. G. (1954). Deposition of salts from sea water by frigid concentration. *Journal of Marine Research*, **13**, 166–82.

Neori, A. & Holm-Hansen, O. (1982). Effect of temperature on rate of photosynthesis in Antarctic phytoplankton. *Polar Biology*, **1**, 33–8.

Neshyba, S. (1977). Upwelling by icebergs. *Nature*, **267**, 507–8.

New Zealand Meteorological Service (1983). Summaries of climatological observations to 1980. *New Zealand Meteorological Service Miscellaneous Publication*, **117**, 172 p.

New Zealand Meteorological Service. (1984). The role of the New Zealand Meteorological Service in antarctic research. *New Zealand Antarctic Record*, **6**, 8–11.

Niebauer, H. J. & Alexander, V. (1985). Oceanographic frontal structure and biological production at an ice edge. *Continental Shelf Research*, **4**, 367–88.

Olson, R. J. (1980). Nitrate and ammonium uptake in Antarctic waters. *Limnology and Oceanography*, **25**, 1064–74.

Ono, T. A. & Murata, N. (1982). Chilling susceptibility of the blue-green alga *Anacystis nidulans*. 3. Lipid phase of cytoplasmic membrane. *Plant Physiology*, **69**, 125–9.

Orchard, V. A. & Corderoy, D. M. (1983). Influence of environmental factors on the decomposition of penguin guano in Antarctica. *Polar Biology*, **1**, 199–204.

Ostrem, G. (1959). Ice melting under a thin layer of moraine and the existence of ice-cores in moraine ridges. *Geografisker Annaler*, **41**, 228–30.

Pace, N. R., Olsen, G. J. & Woese, C. R. (1986). Ribosomal RNA phylogeny and the primary lines of evolutionary descent. *Cell*, **45**, 325–6.

Paige, R. A. (1968). Sub-surface melt pools in the McMurdo Ice Shelf, Antarctica. *Journal of Glaciology*, **7**, 511–16.

Palmisano, A. C., Kottmeier, S. T. & Sullivan, C. W. (1985a). Sea ice microbial communities. IV. The effect of light perturbation on microalgae at the ice–seawater interface in McMurdo Sound. *Marine Ecology – Progress Series*, **21**, 37–45.

Palmisano, A. C., Soo Hoo, J. B., Moe, R. L. & Sullivan, C. W. (1987a). Sea ice microbial communities VII. Changes in under-ice spectral irradiance during the development of Antarctic sea ice microalgal communities. *Marine Ecology – Progress Series*, **35**, 165–73.

Palmisano, A. C., Soo Hoo, J. B., Soo Hoo, S. L., Kottmeier, S. T., Craft, L. L. & Sullivan, C. W. (1986). Photoadaptation in *Phaeocystis pouchetii* advected beneath the annual sea ice in McMurdo Sound, Antarctica. *Journal of Plankton Research*, **8**, 891–906.

Palmisano, A. C., Soo Hoo, J. B. & Sullivan, C. W. (1985b). Photosynthesis–irradiance relationships in sea ice microalgae from McMurdo Sound, Antarctica. *Journal of Phycology*, **21**, 341–6.

Palmisano, A. C., Soo Hoo, J. B. & Sullivan, C. W. (1987b). Effects of four environmental variables on photosynthesis–irradiance relationships in Antarctic sea-ice microalgae. *Marine Biology* **94**, 299–306.

Palmisano, A. C., Soo Hoo, J. B., White, D. C., Smith, G. A., Stanton, G. R. & Burckle, L. H. (1985c). Shade adapted benthic diatoms beneath annual antarctic sea ice. *Journal of Phycology*, **21**, 664–7.

Palmisano, A. C. & Sullivan, C. W. (1982). Physiology of sea ice diatoms. I. Response of three polar diatoms to a simulated summer-winter transition. *Journal of Phycology*, **18**, 489–98.

Palmisano, A. C. & Sullivan, C. W. (1983a). Physiology of sea ice diatoms. II. Dark survival of three polar diatoms. *Canadian Journal of Microbiology*, **29**, 157–60.

Palmisano, A. C. & Sullivan, C. W. (1983b). Antarctic sea ice microbial communities (SIMCO). I. Distribution, diversity and production of ice microalgae in McMurdo Sound, Antarctica in 1980. *Polar Biology*, **2**, 171–7.

Palmisano, A. C. & Sullivan, C. W. (1985a). Physiological response of microalgae in the ice–platelet layer to low-light conditions. In *Antarctic Nutrient Cycles and Food Webs*, ed. W. R. Siegfried, P. R. Condy & R. M. Laws, pp. 84–8. Berlin: Springer-Verlag.

Palmisano, A. C. & Sullivan, C. W. (1985b). Pathways of photosynthetic carbon assimilation in sea-ice microalgae from McMurdo Sound, Antarctica. *Limnology and Oceanography*, **30**, 674–8.

Parker, B. C., Boyer, S., Allnutt, F. C. T., Seaburg, K. G., Wharton, R. A. & Simmons, G. M. (1982a). Soils from the Pensacola Mountains, Antarctica: physical, chemical and biological characteristics. *Soil Biology and Biochemistry*, **14**, 265–71.

Parker, B. C. & Howard, R. V. (1977). The first environmental impact monitoring and assessment in Antarctica. The Dry Valley drilling project. *Biological Conservation*, **12**, 163–77.

Parker, B. C., Keiskell, L. E., Thompson, W. J. & Zeller, E. J. (1978). Non-biogenic fixed nitrogen in Antarctica and some ecological implications. *Nature*, **271**, 651–2.

Parker, B. C., Simmons, G. M., Seaburg, K. G., Cathey, D. D. & Allnutt, F. C. T. (1982b). Comparative ecology of plankton communities in seven Antarctic oasis lakes. *Journal of Plankton Research*, **4**, 271–86.

Parker, B. C., Simmons, G. M., Wharton, R. A., Seaburg, K. G. & Love, F. G. (1982c). Removal of organic and inorganic matter from antarctic lakes by aerial escape of bluegreen algal mats. *Journal of Phycology*, **18**, 72–8.

Parker, B. C. & Wharton, R. A. (1985). Physiological ecology of bluegreen algal mats (modern stromatolites) in Antarctic oasis lakes. *Archiv für Hydrobiologie Supplement*, **71**, 331–48.

Parker, B. C. & Zeller, E. J. (1979). Nitrogenous chemical composition of antarctic ice and snow. *Antarctic Journal of the United States*, **14**, 80–2.

Parkinson, C. L. & Cavalieri, D. J. (1982). Interannual sea ice variations and sea ice/atmosphere interactions in the Southern Ocean 1973–1975. *Annals of Glaciology*, **3**, 249–54.

Parkinson, A. J., Muchmore, H. G., Scott, E. N. & Scott, L. V. (1983). Survival of human parainfluenza viruses in the South Polar environment. *Applied and Environmental Microbiology*, **46**, 901–5.

Parsons, T. R. & Takahashi, M. (1973). *Biological Oceanographic Processes*. Oxford: Pergamon Press Ltd.

Paterson, R. A. & Knox, J. S. (1972). The occurrence of aquatic fungi in Victoria Land and Ross Island. In *Proceedings of the Colloquium on Conservation Problems in Antarctica*, B. C. Parker ed., pp. 185–92. Lawrence, Kansas: Allen Press Inc.

Phillpot, H. R. (1967). *Selected Surface Climatic Data for Antarctic Stations*. Melbourne: Bureau of Meteorology, Australia.

Potts, M. & Bowman, M. A. (1985). Sensitivity of *Nostoc commune* UTEX 584 (Cyanobacteria) to water stress. *Archives for Microbiology*, **141**, 51–6.

Potts, M. & Morrison, N. S. (1986). Shifts in the intracellular ATP pools of immobilised *Nostoc* cells (Cyanobacteria) induced by water stress. *Plant and Soil*, **90**, 211–21.

Priddle, J., Hawes, I., Ellis-Evans, J. C. & Smith, T. J. (1986). Antarctic aquatic ecosystems as habitats for phytoplankton. *Biological Reviews*, **61**, 199–238.

Probyn, T. A. & Painting, S. J. (1985). Nitrogen uptake by size-fractionated phytoplankton populations in Antarctic surface waters. *Limnology and Oceanography*, **30**, 1327–32.

Pugh, G. J. F. & Allsopp, D. (1982). Micro-fungi on Signy Island, South Orkney Islands. *British Antarctic Survey Bulletin*, **57**, 55–67.

Quinn, P. J. (1985). A lipid-phase separation model of low-temperature damage to biological membranes. *Cryobiology*, **22**, 128–46.

Ragotzkie, R. A. & Likens, G. E. (1964). The heat balance of two antarctic lakes. *Limnology and Oceanography*, **9**, 412–25.

Ramsay, A. J. (1983). Bacterial biomass in ornithogenic soils of Antarctica. *Polar Biology*, **1**, 221–5.

Raven, J. A. (1984). A cost-benefit analysis of photon absorption by photosynthetic unicells. *New Phytologist*, **93**, 593–625.

Rawley, B. A. (1982). Diurnal changes in the heterotrophic uptake of glycolate and glucose in two lakes. M.Sc. Thesis, University of Waikato, Hamilton, New Zealand.

Reichardt, W. & Dieckmann, G. (1985). Kinetics and trophic role of bacterial degradation of macro-algae in antarctic coastal waters. In *Antarctic Nutrient Cycles and Food Webs*, eds. W. R. Siegfried, P. R. Condy & R. M. Laws, pp. 115–22. Berlin: Springer-Verlag.

Reynolds, J. M. (1981a). Lakes on George VI Ice Shelf, Antarctica. *Polar Record*, **20**, 425–32.

Reynolds, J. M. (1981b). The distribution of mean annual temperatures in the Antarctica Peninsula. *British Antarctic Survey Bulletin*, **54**, 123–33.

Richardson, M. G. & Whitaker, T. M. (1979). An Antarctic fast-ice food chain: observations on the interaction of the amphipod *Pontogeneia antarctica* Chevreux with ice-associated micro-algae. *British Antarctic Survey Bulletin*, **47**, 107–15.

Rigler, F. H. (1978). Limnology in the high Arctic: a case study of Char Lake. *Verhandlungen der Internationale Vereinigung für Theoretische und Angewandte Limnologie*, **20**, 127–40.

Robarts, R. D. & Zohary, T. (1987). Temperature effects on photosynthetic capacity, respiration and growth rates of bloom-forming cyanobacteria. In *Forum on Cyanobacterial Dominance*, ed. W. F. Vincent, pp. 391–9. *New Zealand Journal of Marine and Freshwater Research*, **21**.

Rochet, M., Legendre, L. & Demers, S. (1985). Acclimation of sea-ice microalgae to freezing temperature. *Marine Ecology – Progress Series*, **24**, 187–91.

Ronner, U., Sorensson, F. & Holm-Hansen, O. (1983). Nitrogen assimilation by phytoplankton in the Scotia Sea. *Polar Biology*, **2**, 137–47.

Ross, R. M. & Quetin, L. B. (1986). How productive are antarctic krill? *BioScience*, **36**, 264–9.

Roulet, N. T. & Adams, W. P. (1984). Illustration of the spatial variability of light entering a lake using an empirical model. *Hydrobiologia*, **109**, 67–74.

Rudolph, E. D. & Benninghoff, W. S. (1977). Competitive and adaptive responses of invading versus indigenous biotas – a plea for organized monitoring. In *Adaptations within Antarctic Ecosystems*, ed. G. A. Llano, pp. 1211–25. Washington DC: Smithsonian Institution.

Russell, N. J. (1984). Mechanisms of thermal adaptation in bacteria: blueprints for survival. *Trends in Biochemical Sciences*, **9**, 108–12.

Sakshaug, E. & Holm-Hansen, O. (1983). Factors governing pelagic production in polar oceans. In *Marine Phytoplankton and Productivity*, ed. O. Holm-Hansen, L. Bolis & R. Gilles, pp. 1–18 Berlin: Springer–Verlag. *Lecture Notes on Coastal and Estuarine Studies*, **8**,

Sakshaug, E. & Holm-Hansen, O. (1986). Photoadaptation in Antarctic phytoplankton: variations in growth rate, chemical composition and P versus I curves. *Journal of Plankton Research*, **8**, 459–73.

Sansom, J. (1984). The temperature record of Scott Base, Antarctica. *New Zealand Journal of Science*, **27**, 21–31.

Savage, M. L. & Stearns, C. (1985). Climate in the vicinity of Ross Island, Antarctica. *Antarctic Journal of the United States*, **20**, 1–9.

Saxena, V. K. (1982). Biogenic nuclei involvement in clouds over the Ross Ice Shelf. *Antarctic Journal of the United States*, **17**, 212–14.

Schlesinger, M. E. (1986). Equilibrium and transient climate warming induced by increased atmospheric $CO_2$. *Climate Dynamics*, **1**, 35–51.

Schlichting, H. E., Speziale, B. J. & Zink, R. M. (1978). Dispersal of algae and protozoa by antarctic flying birds. *Antarctic Journal of the United States*, **13**, 147–9.

Schmetterer, G. & Peschek, G. A. (1981). Treatments effecting reversible photobleaching and thylakoid degradation in the blue-green alga *Anacystis nidulans*. *Biochemica Physiologia Plantarum*, **176**, 90–100.

Schwerdtfeger, W. (1983). The climate of the Antarctic. In *Climates of the Polar Regions*, ed. S. Orvig, pp. 253–331. Amsterdam–London: Elsevier.

Schwerdtfeger, W. (1984). *Weather and Climate of the Antarctic. Developments in Atmospheric Science, 15*, New York: Elsevier.

Seaburg, K. G., Parker, B. C., Prescott, G. W. & Whitford, L. A. (1979). *The Algae of Southern Victorialand, Antarctica. Bibliothecia Phycologia, 46*. Vaduz: J. Kramer.

Seaburg, K. G., Parker, B. C., Wharton, R. A. & Simmons, G. M. (1981). Temperature-growth response of algal isolates from Antarctic oases. *Journal of Phycology*, **17**, 353–360.

Shephard, K. L. (1987). Evaporation of water from the mucilage of a gelatinous algal community. *British Phycological Journal*, **22**, 181–5.

Silver, M. W., Mitchell, J. G. & Ringo, D. L. (1980). Siliceous nanoplankton. II. Newly discovered cysts and abundant choanoflagellates from the Weddell Sea, Antarctica. *Marine Biology*, **58**, 211–17.

Singer, J. (1982). Role of benthic organisms in sediment erosion and transport. *Antarctic Journal of the United States*, **17**, 118–19.

Slawyk, G. (1979). $^{13}$C and $^{15}$N uptake by phytoplankton in the Antarctic upwelling area: results from the Antiprod I cruise in the Indian Ocean sector. *Australian Journal of Marine and Freshwater Research*, **30**, 431–48.

Smith, D., Coulson, G. E. & Morris, G. J. (1986). A comparative study of the morphology and viability of hyphae of *Penicillium expansum* and *Phytophthora nicotianae* during freezing and thawing. *Journal of General Microbiology*, **132**, 2013–21.

Smith, G. A., Nichols, P. D. & White, D. C. (1988). Fatty acid composition and microbial activity of benthic marine sediment from McMurdo Sound, Antarctica. *FEMS Microbiology Ecology*, in press.

Smith, G. A., White, D. C. & Nichols, P. D. (1986). Antarctic benthic and sea-ice microalgal interactions: food chain processes and physiology. *Antarctic Journal of the United States*.

Smith, H. G. (1978). The distribution and ecology of terrestrial protozoa of sub-Antarctic and maritime Antarctic islands. *British Antarctic Survey Scientific Reports*, **95**, 104 p.

Smith, H. G. (1982). The terrestrial protozoan fauna of South Georgia. *Polar Biology*, **1**, 173–9.

Smith, H. G. (1984). The protozoa of Signy Island fellfields. *British Antarctic Survey Bulletin*, **64**, 55–61.

Smith, H. G. & Tearle, P. V. (1985). Aspects of microbial and protozoan abundances in Signy Island fellfields. *British Antarctic Survey Bulletin*, **68**, 83–90.

Smith, J. C. & Platt, T. (1985). Temperature responses of ribulose bisphosphate carboxylase and photosynthetic capacity in arctic and tropical phytoplankton. *Marine Ecology – Progress Series*, **25**, 31–7.

Smith, V. R. & Russell, S. (1982). Acetylene reduction by bryophyte–cyanobacteria associations on a subantarctic island. *Polar Biology*, **1**, 153–7.

Smith, W. O. & Nelson, D. M. (1985). Phytoplankton bloom produced by a receding ice edge in the Ross Sea: Spatial coherence with the density field. *Science*, **227**, 163–6.

Solopov, A. V. (1969). *Oases in Antarctica*. Jerusalem: Israel Programme for Scientific Translations.

Sommer, U. (1986). Nitrate- and silicate-competition among antarctic phytoplankton. *Marine Biology*, **91**, 345–51.

Sorokin, Ju. I. (1971). On the role of bacteria in the productivity of tropical oceanic waters. *Internationale Revue der gesamten Hydrobiologie*, **56**, 1–48.

Spaull, V. W. (1973). Distribution of nematode feeding groups at Signy Island, South Orkney Islands, with an estimate of their biomass and oxygen consumption. *British Antarctic Survey Bulletin*, **37**, 21–32.

Speir, T. W. & Cowling, J. C. (1984). Ornithogenic soils of the Cape Bird Adélie penguin rookeries Antarctica. I. Chemical properties. *Polar Biology*, **2**, 199–205.

Speir, T. W. & Ross, D. J. (1984). Ornithogenic soils of the Cape Bird Adélie penguin rookeries, Antarctica. 2. Ammonia evolution and enzyme activities. *Polar Biology*, **2**, 207–12.

Squire, V. A. (1984). Sea ice. *Science Progress, Oxford*, **69**, 19–43.

Steponkus, P. L. (1981). Responses to extreme temperatures. Cellular and sub-cellular bases. In *Physiological Plant Ecology I Responses to the Physical Environment*, ed. Lange, O. L., Nobel, P. S., Osmund, C. B. & Ziegler, H., pp. 371–402. Berlin: Springer-Verlag.

Sugden, D. (1982). *Arctic and Antarctic. A Modern Geographical Synthesis*. Oxford: Basil Blackwell Publisher Ltd.

Sullivan, C. W. (1985). Sea ice bacteria: reciprocal interactions of the organisms and their environment. In *Sea Ice Biota*, ed. R. Horner, pp. 159–72. Florida: CRC Press.

Sullivan, C. W. & Palmisano, A. C. (1981). Sea ice microbial communities in McMurdo Sound. *Antarctic Journal of the United States*, **16**, 126–7.

Sullivan, C. W. & Palmisano, A. C. (1984). Sea ice microbial communities (SIMCO): Distribution, abundance and diversity of ice bacteria in McMurdo Sound, Antarctica, in 1980. *Applied and Environmental Microbiology*, **47**, 788–95.

Sullivan, C. W., Palmisano, A. C., Kottmeier, S. T., McGrath Grossi, S., Moe, R. & Taylor, G. T. (1983). The influence of light on development and growth of sea-ice microbial communities in McMurdo Sound. *Antarctic Journal of the United States*, **18**, 177–9.

Swithinbank, C. (1970). Ice movement in the McMurdo Sound area of Antarctica. In pp. 472–86. *Proceedings of the International Symposium on Antarctic Glaciological Exploration*.

Tanimura, A., Minoda, T., Fukuchi, M., Hoshiai, T. & Ohtsuka, H. (1984). Swarm of *Paralabidocera antarctica* (Calanoida, Copepoda) under sea ice near Syowa Station, Antarctica. *Antarctic Record (Japan)*, **92**, 12–19.

Tanner, A. C. (1985). The role of bacteria in the cycling of nutrients within the maritime Antarctic environment. In *Antarctic Nutrient Cycles and Food Webs*, ed. W. R. Siegfried, P. R. Condy & R. M. Laws, pp. 123–7. Berlin: Springer-Verlag.

Tanoue, E. & Hara, S. (1986). Ecological implications of fecal pellets produced by the Antarctic krill *Euphausia superba* in the Antarctic Ocean. *Marine Biology*, **91**, 359–69.

Tchernia, P. (1980). *Descriptive Regional Oceanography. Pergamon Marine Series, 3*. New York: Pergamon Press.

Thompson, D. C., Craig, R. M. F. & Bromley, A. M. (1971). Climate and surface heat balance in an Antarctic dry valley. *New Zealand Journal of Science*, **14**, 245–51.

Thompson, T. G. & Nelson, K. H. (1956). Concentration of brines and deposition of salts from sea water under frigid conditions. *American Journal of Science*, **254**, 227–38.

Tilzer, M. M. & Dubinsky, Z. (1987). Effects of temperature and day length on the mass balance of Antarctic phytoplankton. *Polar Biology*, **7**, 35–42.

Tilzer, M. M., Elbrachter, M., Gieskes, W. W. & Beese, B. (1986). Light-temperature interactions in the control of photosynthesis in Antarctic phytoplankton. *Polar Biology*, **5**, 105–11.

Tilzer, M. M., Von Bodungen, B. & Smetacek, V. (1985). Light-dependence of phytoplankton photosynthesis in the Antarctic Ocean: implications for regulating productivity. In *Antarctic Nutrient Cycles and Food Webs*, ed. W. R. Siegfried, P. R. Condy, & R. M. Laws, pp. 60–9. Berlin: Springer-Verlag.

Torii, T. (1975). *Geochemical and Geophysical Studies of Dry Valleys, Victoria Land in Antarctica. Memoirs of the National Institute of Polar Research, Special Issue, 4*. Tokyo: National Institute of Polar Research.

Torii, T. & Ossaka, J. (1965). Antarcticite: a new mineral, calcium chloride hexahydrate, discovered in Antarctica. *Science*, **149**, 975–7.

Torres, J. J., Lancraft, T. M., Weigle, B. L., Hopkins, T. L. & Robison, B. H. (1984). Distribution and abundance of fishes and salps in relation to the marginal ice zone of the Scotia Sea, November and December 1983. *Antarctic Journal of the United States*, **19**, 117–19.

Tranter, D. J. (1982). Interlinking of physical and biological processes in the Antarctic Ocean. *Oceanography Marine Biology Annual Review*, **20**, 11–35.

Treshnikov, A. F. (1966). The ice of the Southern Ocean. *Japanese Antarctic Research Expedition Scientific Reports, Special Issue No. 1*, 113–23.

Tschermak-Woess, E. & Friedmann, E. I. (1984). *Hemichloris antarctica* gen. et sp. nov. (Chlorococcales, Chlorophyta), a cryptoendolithic alga from Antarctica. *Phycologia*, **23**, 443–54.

Ugolini, F. C. (1972). Ornithogenic soils of Antarctica. In *Antarctic Terrestrial Biology. Antarctic Research series, 20*. ed. G. A. Llano, 181–93. Washington DC: American Geophysical Union.

Ugolini, F. C. & Starkey, R. L. (1966). Soils and micro-organisms from Mt Erebus, Antarctica. *Nature*, **211**, 440–1.

van den Hoff, J. & Franzmann, P. (1986). A choanoflagellate in a hypersaline antarctic lake. *Polar Biology*, **6**, 71–3.

Vestal, J. R. (1985). Effects of light intensity on the cryptoendolithic microbiota. *Antarctic Journal of the United States*, **20**, 181–2.

Vestal, J. R. & Friedmann, E. I. (1982). *In situ* carbon metabolism by the cryptoendolithic microbial community in the antarctic cold desert. *Antarctic Journal of the United States*, **17**, 190–1.

Vincent, W. F. (1981). Production strategies in antarctic inland waters: phytoplankton eco-physiology in a permanently ice-covered lake. *Ecology*, **62**, 1215–24.

Vincent, W. F. (1987). Extreme aquatic environments II. Antarctic limnology. In *Inland Waters of New Zealand*, ed. A. B. Viner, pp. 379–412. Wellington; Science Information Publishing Centre.

Vincent, W. F., Downes, M. T. & Vincent, C. L. (1981). Nitrous oxide cycling in Lake Vanda, Antarctica. *Nature*, **292**, 618–20.

Vincent, W. F. & Howard-Williams, C. (1986). Antarctic stream ecosystems: physiological ecology of a blue-green algal epilithon. *Freshwater Biology*, **16**, 219–33.

Vincent, W. F. & Howard-Williams, C. (1987). Microbial ecology of antarctic streams. In *Proceedings of the Fourth International Congress on Microbial Ecology*, ed. F. Megusar. In press.

Vincent, W. F. & Howard-Williams, C. (1988). Microbial communities in southern Victoria Land streams (Antarctica). II. The effects of low temperature. In *High Latitude Limnology*, ed. W. F. Vincent & J. C. Ellis-Evans. The Hague: Junk Publishers. In press.

Vincent, W. F. & Vincent, C. L. (1982a). Factors controlling phytoplankton production in Lake Vanda (77 °S). *Canadian Journal of Fisheries and Aquatic Sciences*, **39**, 1602–9.

Vincent, W. F. & Vincent, C. L. (1982b). Response to nutrient enrichment by the plankton of Antarctic coastal lakes and inshore Ross Sea. *Polar Biology*, **1**, 159–65.

Vishniac, H. S. (1983). An enation system for the isolation of Antarctic yeasts inhibited by conventional media. *Canadian Journal of Microbiology*, **29**, 90–5.

Vishniac, H. S. (1985). *Cryptococcus friedmannii*, a new species of yeast from the Antarctic. *Mycologia*, **77**, 149–53.

Vishniac, H. S. & Baharaeen, S. (1982). Five new basidioblastomycetous yeast species segregated from *Cryptococcus vishniacii* emend. auct., an antarctic yeast species comprising four new varieties. *International Journal of Systematic Bacteriology*, **32**, 437–45.

Vishniac, H. S. & Hempfling, W. P. (1979). Evidence of an indigenous microbiota (yeast) in the dry valleys of Antarctica. *Journal of General Microbiology*, **112**, 301–14.

Vollenweider, R. A., Munawar, M. & Stadelmann, P. (1974). A comparative review of phytoplankton and primary production in the Laurentian Great Lakes. *Journal of the Fisheries Research Board of Canada*, **31**, 739–62.

Wada, E., Shibata, R. & Torii, T. (1981). [15]N abundance in Antarctica: origin of soil nitrogen and ecological implications. *Nature*, **292**, 327–9.

Wager, A. C. (1972). Flooding of the ice shelf in George VI Sound. *British Antarctic Survey Bulletin*, **28**, 71–4.

Walsh, J. E. & Johnson, C. M. (1979). An analysis of Arctic sea ice fluctuations, 1953–77. *Journal of Physical Oceanography*, **9**, 580–91.

Walton, D. W. H. (1977). Radiation and soil temperature 1972–74: Signy Island terrestrial reference site. *British Antarctic Survey Data*, **1**, 1–51.

Walton, D. W. H. (1982). The Signy Island reference sites: XIII. Microclimate monitoring, 1972–74. *British Antarctic Survey Bulletin*, **55**, 111–26.

Walton, D. W. H. (1984). The terrestrial environment. In *Antarctic Ecology*, ed. R. M. Laws, pp. 1–60. London: Academic Press.

Ward, B. L. (1984). Distribution of modern benthic foraminifera of McMurdo Sound, Antarctica. Unpublished Ph.D. thesis, Victoria University of Wellington, New Zealand, 211 p.

Ward, D. M. & Winfrey, M. R. (1985). Interactions between methanogenic and sulfate-reducing bacteria in sediments. *Advances in Aquatic Microbiology*, 141–79.

Warren Wilson, J. (1958). Dirt on snow patches. *Journal of Ecology*, **46**, 191–8.

Watanuki, T. (1979). Isolation and culture of antarctic diatoms from the saline lakes in the Soya Coast, East Antarctica. *Memoirs of the National Institute of Polar Research, Special issue 11*, 35–41.

Weeks, W. F. & Ackley, S. F. (1982). The growth, structure, and properties of sea ice. *CCREL Monograph 82–1*. Hanover, New Hampshire: United States Army Cold Regions Research and Engineering Laboratory.

Weeks, W. F. & Lee, O. S. (1962). The salinity distribution in young sea ice. *Arctic*, **15**, 92–109.

Weiss, R. F. (1970). The solubility of nitrogen, oxygen and argon in water and seawater. *Deep-Sea Research*, **17**, 721–35.

Wharton, R. A., McKay, C. P., Mancinelli, R. L. & Simmons, G. M. (1987). Perennial $N_2$ supersaturation in an antarctic lake. *Nature*, **325**, 343–5.

Wharton, R. A., McKay, C. P., Simmons, G. M. & Parker, B. C. (1985). Cryoconite holes on glaciers. *BioScience*, **35**, 499–503.

Wharton, R. A., McKay, C. P., Simmons, G. M. & Parker, B. C. (1986). Oxygen budget of a perennially ice-covered Antarctic lake. *Limnology and Oceanography*, **31**, 437–43.

Wharton, R. A., Parker, B. C. & Simmons, G. M. (1983). Distribution, species composition and morphology of algal mats (stromatolites) in Antarctic dry valley lakes. *Phycologia*, **22**, 355–65.

Wharton, R. A., Vinyard, W. C., Parker, B. C., Simmons, G. M. & Seaburg, K. G. (1981). Algae in cryoconite holes on Canada Glacier in southern Victoria Land, Antarctica. *Phycologia*, **20**, 208–11.

Whitaker, T. M. (1977). Sea ice habitats of Signy Island (South Orkneys) and their primary productivity. In *Adaptations within Antarctic Ecosystems*, ed. G. A. Llano, pp. 75–83. Washington DC: Smithsonian Institution.

White, D. C., Smith, G. A. & Stanton, G. R. (1984). Biomass, community structure, and metabolic activity of the microbiota in benthic marine sediments and sponge spicules mats. *Antarctic Journal of the United States*, **19**, 125–6.

White, M. G. (1984). Marine benthos. In *Antarctic Ecology*, ed. R. M. Laws, pp. 421–61. London: Academic Press.

Wiebe, W. J. & Hendricks, C. W. (1974). Distribution of heterotrophic bacteria in the transect of the Antarctic Ocean. In *Effect of the Ocean Environment on Microbial Activity*, ed. R. R. Colwell & R. Y. Morita, pp. 524–35. Baltimore: University Park Press.

Wilson, A. T. (1965). Escape of algae from frozen lakes and ponds. *Ecology*, **46**, 376.

Wilson, D. L., Smith, W. O. & Nelson, D. M. (1986). Phytoplankton bloom dynamics of the western Ross Sea ice edge-I. Primary productivity and species-specific production. *Deep-Sea Research*, **33**, 1375–87.

Woods, J. L. (1976). Aspects of the biology of the Antarctic freshwater rotifer *Philodina gregaria*. M.Sc. thesis, University of Canterbury, New Zealand.

Wrenn, J. H. & Beckman, S. W. (1981). Maceral and total organic carbon analyses of DVDP drill core II. *Antarctic Research Series*, **33**, 391–402.

Wright, S. W. & Burton, H. R. (1981). The biology of Antarctic saline lakes. *Hydrobiologia*, **82**, 319–38.

Wynn-Williams, D. D. (1980). Seasonal fluctuations in microbial action in Antarctic moss peat. *Biological Journal of the Linnean Society*, **14**, 11–28.

Wynn-Williams, D. D. (1983). Distribution and characteristics of *Chromobacterium* in the Maritime and Sub-antarctic. *Polar Biology*, **2**, 101–8.

Wynn-Williams, D. D. (1984). Comparative respirometry of peat decomposition on a latitudinal transect in the maritime antarctic. *Polar Biology*, **3**, 173–81.

Wynn-Williams, D. D. (1985). Photofading retardant for epifluorescence microscopy in soil micro-ecological studies. *Soil Biology and Biochemistry*, **17**, 739–46.

Wynn-Williams, D. D. (1987). Microbial colonization of antarctic fellfield soils. In *International Congress on Microbial Ecology*, ed. F. Megusar. Ljubljana.

Yarrington, M. R. & Wynn-Williams, D. D. (1985). Methanogenesis and the anaerobic microbiology of a wet moss community at Signy Island. In *Antarctic Nutrient Cycles and Food Webs*, ed. W. R. Siegfried, P. R. Condy & R. M. Laws, pp. 229–33. Berlin: Springer-Verlag.

Zeller, E. J. & Parker, B. C. (1981). Nitrate ion in antarctic firn as a marker for solar activity. *Geophysical Research Letters*, **8**, 895–8.

Zunino, C. H., Castrelos, O. D. & Margni, R. A. (1985). Antarctic microbiology. IV. Contributions for its better understanding in Puerto Paraiso. *Instituto Antartico Argentino Contribucion*, 307.

Zwally, H. J., Comiso, J. C. & Gordon, A. L. (1985). Antarctic offshore leads and polynyas and oceanographic effects. In *Oceanology of the Antarctic Continental Shelf. Antarctic Research Series, 43*, ed. S. S. Jacobs, pp. 203–26. Washington DC: American Geophysical Union.

Zwally, H. J., Parkinson, C., Carsey, F., Gloersen, P., Campbell, W. J. & Ramseier, R. O. (1979). Antarctic sea ice variations 1973–75. *NASA Weather Climate Review*, **56**, 335–40.

Zwally, H. J., Parkinson, K. L. & Comiso, J. C. (1983). Variability of Antarctic sea ice and changes in carbon dioxide. *Science*, **220**, 1005–12.

# Index

Printed in the United States
by Baker & Taylor Publisher Services

Printed in the United States
By Bookmasters